Communication, Culture and Change in Asia

Volume 3

Series editor

Jan Servaes, Hong Kong, China

Patchanee Malikhao

Culture and Communication in Thailand

 Springer

Patchanee Malikhao
Fecund Communication
Chiangmai
Thailand

ISSN 2366-4665   ISSN 2366-4673   (electronic)
Communication, Culture and Change in Asia
ISBN 978-981-10-4123-5   ISBN 978-981-10-4125-9   (eBook)
DOI 10.1007/978-981-10-4125-9

Library of Congress Control Number: 2017934870

Printed on acid-free paper

This Springer imprint is published by Springer Nature
The registered company is Springer Nature Singapore Pte Ltd.
The registered company address is: 152 Beach Road, #21-01/04 Gateway East, Singapore 189721, Singapore

The original version of this book was revised.
An erratum to this book can be found at
DOI 10.1007/978-981-10-4125-9_10

To Jan, Fiona, and Lisa, my family
And to all people in search of a humane and
sustainable future,
Especially those living in Thailand

Acknowledgements

Trotting around the globe and finishing this book at the same time made its completion an unforgettable experience!

I would like to thank Jan Servaes, Fiona Servaes, and Lisa Servaes for their contribution. It shows their deep interest in Thai culture, environment, and future.

I would like to thank Philip Lee, editor of Media Development Journal for allowing me to use a modified and updated version of Malikhao, P. (2013). Communication for Development and Sustainability. *Media Development*, 3, pp. 1–5 as part of the introduction of this book.

I would also like to thank Father Franz-Jozef Eilers, editor of the *Journal of the Asian Research Center for Religion Communication* for giving me permission to use Malikhao, P. (2015). Thai Buddhism, the Mass Media and Culture Change in Thailand. In *Journal of the Asian Research Center for Religion Communication*. 12(2), pp. 124–143 as Chap. 1 in this book.

I am grateful that Springer Publishers wanted to publish this book.

Contents

1 **Thai Buddhism, the Mass Media, and Culture Change**
   in Thailand 1
   1.1   Introduction 1
   1.2   Thai Buddhism in the Phases of Globalization 2
   1.3   Thai Culture: A Mediatized Buddhist Culture. 8
   1.4   Conclusion 13
   References. 13

2 **Analyzing the "Dhammakaya Case" Online** 17
   2.1   Introduction 17
   2.2   Objectives of This Study 20
   2.3   Research Questions. 21
   2.4   Methodology 21
   2.5   Findings and Discussion. 21
   2.6   Analysis 25
   2.7   The 12 Causal Links in the PS Model 27
   2.8   Analysis 28
   2.9   Analysis 29
   2.10   Conclusion 33
   References. 33

3 **Violence Against Thai Females**. 37
   3.1   Introduction 37
   3.2   Female Status in Thai Culture from a Historical Perspective 38
   3.3   Analysis: Gender and Power as Represented in the Media. 43
   3.4   Conclusion 45
   References. 46

4   A Village in the Jungle: Culture and Communication
in Thailand   49
   4.1   Introduction   49
   4.2   Thai Feudalism: The "Sakdina" System   51
   4.3   A Rural Village Culture   52
   4.4   The Power of Beliefs   54
   4.5   The Amoral Power ... and Moral Kindness   56
   4.6   Interpersonal Communication: Mai Pen Rai   56
   4.7   Likay Drama   59
   4.8   Growing Pains: Modernization and Tradition   59
   4.9   The Thai Value System   61
   4.10   Thai Leaders and Face   64
   4.11   Conclusion   65
   References   66

5   Tourism, Digital Social Communication and Development
Discourse: A Case Study on Chinese Tourists in Thailand   71
   5.1   Introduction   71
   5.2   Literature Review and Theoretical Perspectives   74
      5.2.1   Globalization and Culture   74
      5.2.2   Globalization and Mediatized Society   74
      5.2.3   Mediatization and Culture   75
      5.2.4   Development Discourse, Sustainability,
           and EcoTourism   77
      5.2.5   Research Questions   78
      5.2.6   Methodology   78
      5.2.7   General Observations   79
      5.2.8   Answering the Research Questions   83
   5.3   Conclusion   85
   References   85

6   Self-Reliance and Sustainability from a Thai Perspective   89
   6.1   Introduction   89
   6.2   Globalization and Sustainable Development   89
   6.3   The Multiplicity Paradigm   93
   6.4   King Bhumibol's Economic Sufficiency   93
   6.5   Self-Reliance and Sustainability from a Thai Perspective:
      Lessons Learned at Punpun Organic Farm, Phrao District,
      Chiang Mai, Thailand   96
      6.5.1   Background of Punpun   96
   6.6   What I Learned from Punpun   97
   6.7   Conclusion   101
   References   102

7   **Mindful Communication and Journalism from a Thai Buddhist**
    Perspective   103
    7.1   Introduction   103
    7.2   How Thai Buddhism Helps Us See Through the Perception
          of Self!   104
    7.3   Not-Self   106
    7.4   Twelve Causally Linked Stages of the Paticcasamuppada   107
    7.5   Mindful Communication   109
    7.6   Mindful Journalism   111
          7.6.1   Selfless Journalists as Actors/Agents/Propellers
                  for Social Interactions and Social Change   112
    7.7   Conclusion   113
    References   114

8   **Human Trafficking in Thailand: A Culture of Corruption**   117
    8.1   Introduction   117
    8.2   Sex Trafficking in Thailand   118
    8.3   Men as Sex Trafficking Victims   119
    8.4   Child Trafficking in Thailand   120
    8.5   Forced Labor in Thailand's Seafood Industry   121
    8.6   TIP Report   121
    8.7   How to End Corruption in Thailand?   122
    References   123

9   **Elephants in Tourism. Sustainable and Practical Approaches**
    to Captive Elephant Welfare and Conservation in Thailand   127
    9.1   Introduction   127
    9.2   An Anthropocentric History   128
    9.3   Welfare Issues   129
    9.4   Elephant Tourism and Conservation   131
    9.5   Plight of the Mahout   132
    9.6   Role of Western NGOs   132
    9.7   Toward Sustainable Solutions   134
    9.8   Concluding Remarks   136
    References   137

Erratum to: Culture and Communication in Thailand E1

Index   139

About the Author

Patchanee Malikhao received her Ph.D. in Sociology from the University of Queensland in Australia, a Master's of Arts degree in Mass Communication from Thammasat University in Thailand, a Master's of Science degree in Printing Technology from Rochester Institute of Technology (RIT), Rochester, New York, and a Bachelor of Science (Hons.) in Photographic Science and Printing Technology from Chulalongkorn University in Thailand. She has worked and received extensive trainings in the fields of Communication for Social Change, Imaging Technology, Social Science Research Methods and Data Analyses in Belgium, Australia, and the United States. She was a recipient of many scholarships and awards, including the Fulbright Scholarship, the Australian Postgraduate Award, and the Outstanding Teacher Award. She had worked as researcher and lecturer at many places in the world: Chulalongkorn University, Thailand; The University of Queensland, Australia; University of Massachusetts Amherst, USA; Thammasat University, Thailand. She is a Senior Fellow in the CSSC Center at UMass Amherst since 2010. She has published in the fields of Development Communication, Health Communication, Religion Communication; Buddhism and Journalism; Sustainable Development; and HIV/AIDS Sociology. Her latest book is "Effective Health Communication for Sustainable Development", published by Nova Science in New York in 2016. An earlier book, *Sex in the Village. Culture, Religion and HIV/AIDS in Thailand*, was published by Southbound & Silkworm Publishers: Penang-Chiang Mai, in 2011.

List of Figures

Fig. 4.1   Main characteristics of Thai culture.   58

List of Tables

Table 2.1　Results of "ธรรมกาย" (Dhammakaya) keyword search on Google (research undertaken on March 27, 2015 and December 26–29, 2015) . . . .　22

Table 2.2　Results of googling #ธัมมี ธรรมกาย (Dhammy Dhammakaya), researched on March 27, 2015 . . . .　23

Table 2.3　Results of googling # นะจ๊ะธรรมกาย (Na ja Dhammakaya); research undertaken on March 27, 2015 . . . .　24

Table 4.1　Summary of the three religious subsystems of Thai religion . . . .　55

Table 4.2　Thai worldviews, values, and symbolic representations . . . .　63

Introduction

When I was a child in Thailand in the 70s, our family often went from Bangkok to visit our uncle who lived upcountry. Year after year, we heard that the village where our uncle had a pig farm and paddy fields became more and more "developed." That meant that he had a better road to his house; he did not need to generate electricity anymore as electricity poles came to every household in the village; he had tap water instead of having to pump water from the irrigation canal to his home. Later, I also learned that my country, Thailand, was called a developing country as the infrastructure needed to be called "developed" had been clustered in the capital city, Bangkok. Moreover, our GDP was so low. Thais who live under the poverty line in 2009 were 8.1% of the population. That is about 5.6 million people who earn at most US\$ 1.25 a day (indexmundi.com). A friend said, "You should be happy that they eulogize our status a bit. We are not underdeveloped. We are developing."

Well, the terms "developed" was meant to be "material growth and how much money you make." According to the website whereisthailand.info, only 2% of the population in Thailand has more than one million Baht and above savings in the bank. Are these 2% of the population developed then? Ironically I witnessed developed Thais who drive a Mercedes-Benz and litter on the street; people with big cars park anywhere they please and obstruct the traffic; not to mention about frauds and scandals created by rich Thais.

From the late 1980s to the beginning of the 2000s, I lived in the Netherlands and Belgium. I observed the conflicts between the hosts (the Europeans) and migrant workers from Turkey and Morocco. Being developed meant keeping the wealth among white Caucasians? My following experience will illuminate more. During 2003–2007, I was in Australia, one of the richest countries in the world. "You need to survive financially for the first two years because Australia is a developed country; the cost of living is expensive"—that's what we read in a leaflet advising people about how to migrate to the country. When we were there, we saw pictures in newspapers depicting aborigines who live faraway in the outback without electricity, they had to go on foot to school, lacked tap water and lived in a reservation area. It was a shock that we lived in a nice neighborhood with electricity, tap water, regular public transportation, internet and cable TV access, gas lines etc. This contrast made me rethink the term "development". In my view, the natives have the right to get a fair share of development they deserve. That is what scholars call, "social justice." Income distribution and a fair share of a nation's "wealth" should be one of the indicators for development too.

Five years later, I lived in the US; in Amherst, Massachusetts, to be precise. I observed shattered, rundown, and deserted factories in many towns during the six years we lived and worked there. The center of mass production had already shifted to other parts of the world where labors are cheaper and the threshold levels of pollution can be lower. I witnessed both African Americans and Caucasians who became homeless, begging for dollars at a few crossroads in our town. Sad news broke daily: mass massacres from shooting sprees in many places; foreclosures of homes; huge deficits as a consequence of the Iraq and Afghanistan wars; and severe budget cuts that affected the education system, airlines, social services and more. Of course, we have electricity, tap water, water closets, internet access, a central heating system, and a phone. It is alarming that 46 millions of Americans still live below the poverty line. In 2012, a single person under 65 years old who earns less than $11,945 per year is considered poor (concensus.gov).

On average, a poor American earns about 33 dollars a day, which is almost 30 times more than a poor Thai. In other words, development is quite relative, isn't it? The level of dissatisfaction in the quality of life may be more or less the same, but the amount of money one earns is so different. Does development mean safety of life, a safety net for the underserved, and income distribution? Can the people in those almost deserted towns get organized and revitalize their own communities?

You see, development is a discourse. It has many different meanings and definitions, as Wolfgang Sachs (2013:5), a German scholar in development dialogue, puts it:

> "On the one hand there are those who implicitly identify development with economic growth, calling for more relative equity in GDP. Their use of the word 'development' reinforces the hegemony of the economic world view... On the other hand, there are those who identify development with more rights and resources for the poor and powerless. Their use of the word calls for de-emphasizing growth in favor of greater autonomy of communities."

Regarding development theories, the modernization paradigm which endorses economic wealth has been popular since the 1960s. It emphasizes development as a path of growth that a poorer country should follow to get into the holy grail of living like Western people. That assumption is being proven to be inadequate. The slumping down of the European and the US economy; the busting of the inflated housing market in Japan and the US; the droughts and strong UV levels that cause skin cancer in Australia; the suicidal cases in Japan and Korea, the pollution of air and water in China ... should not be dream goals for any developing country.

Now, it is clear that development does not mean that a Thai farmer should aim to earn and spend like an American farmer. Westernization should not be the ultimate goal of development, said Jan Servaes (1999) in his famous, "One World. Multiple Cultures" book. The booming of economies at the expense of environmental disasters or at the global climate change level should not be considered as desirable development goals. They are definitely *not* sustainable development goals.

The People's Republic of China is a case in point. Recently, the Chinese government admitted that there are a number of cancer villages where water is no longer drinkable due to chemical waste dumps in water ways. Smog from factory smoke stacks caused respiratory problems too (thescientist.com). At present, citizens in Beijing and other major cities in China, and in other developing countries, are suffering from more than threshold levels of pollution in the name of modernization and growth. Instead of battling poverty, diseases and ignorance, governments of developing countries implemented "modern" economic and environmental policies to aggravate poverty-stricken grassroots and induce new sorts of cancer and respiratory diseases out of ignorance.

The dependistas in Latin America in the 1980s proposed a new paradigm called, "Dependency Paradigm". Criticizing multi-national corporations and neo-colonialism that moved natural and human resources around the globe in the name of "globalization." The dependistas were criticized as well, as they put too much blame on outside actors and less on their own problem of internal colonization in which the wealthier elites are exploiting the poor in one's own country. Many governments seem to ignore these lessons as well, including consecutive Thai governments.

Now let's move to the alternative paradigm: the Multiplicity Paradigm, proposed by Jan Servaes in the 1990s. This paradigm emphasizes participatory communication and democratic planning strategies to achieve one's desired goal of development. Development goes hand in hand with social change and that must be sustained too. If it is not sustainable development, our natural resources will be exhausted and the environmental impacts, such as green house effects, climate change, the dislocation of plates due to the melting of the ice in the North Pole, the big hole in the ozone layer in the Southern hemisphere, etc. will drive our planet into extinction.

Development cannot be achieved without sustainable social change! That is why the Bhutanese government announced in 2010 that Bhutan is not going to consider just the growth of GDP to be the only indicator for its development. The Bhutanese use the term GNH or gross national happiness. Though, it is difficult to fathom the level of happiness, it is obvious from a number of examples given above that having more money does not mean having more happiness. The Bhutanese GNH can be defined as good governance, sustainable socio-economic development, cultural preservation, and environmental conservation. They are applying the Multiplicity paradigm in their framework of sustainable development, it seems. This is more congruent to the kind of development we want. We don't want to have cancer. We don't want epidemics. We want a clean and green environment, clean water and food, good education for our children, good medication and healthcare, good elderly care and childcare. In sum, we all want peace and happiness. We want all good things.

But why do we want others to get what we don't want? It is now proven that the world is related and interrelated. What we don't want may come to us inevitably by air, water ways, and physical connections. There is a saying in Chinese that the reverberating energy of picking a flower can be felt among the stars. That means, we should see and foresee the relations of beings and nature in a holistic way. Having a good attitude toward the natural environment and conserve one's own natural resources is what we call 'being educated'. Being uneducated, in my view, means seeing things in fragments, aiming at exploiting other humans and nature at an individual level or at a national level, considering only monetary profits, having a fixed mindset/worldview/beliefs, and not being able to appreciate others who are different in cultures, social status, education, race, and ethnicity. Therefore, we urgently need to re-educate ourselves and plan to properly educate the new generations. Development from within is needed as part of a moral ethics curriculum to promote national pride. It should be passed on from one generation to the next.

Now it comes to communication to enable social change for the better. Good governance, civic society, participatory communication and all cannot be achieved without good communication, both mass and interpersonal through old and new (social) media. Servaes (2013: 317) proposes three streams of action: first, the media must be activated to build up advocacy for policy decisions; second, networking among interest groups and alliances, individuals, political forces, academic and non-academic organizations, business, industry, religious groups etc. is needed; public demand and movements of citizens to push development issues and agendas are needed. Participation and power in its nature and kind are analyzed in detail in another of Jan Servaes' books, "Sustainability, Participation and Culture in Communication", published in 2013. What I would like to add here is that we are dealing with different groups and different subcultures; therefore, intercultural communication is needed as a pre-requisite to advocacy participatory communication. Martin and Nakayama (2004: 62) recommend the dialectical approach to intercultural communication: first, it is important to remember that cultures change as do individuals; second, one should view various aspects of culture in a holistic perspective; and third, holding contradictory ideas simultaneously is part of life. They advise to consider both group culture and individual characteristics; the communication context; cultural relativity, the realization that culture is dynamic; history, present and future; and understanding that all of us have some privilege and disadvantage in one way or the other.

We need to realize that participation is not just a lip-service. Participation under the framework of Multiplicity recognizes felt needs, environmental concerns, self-reliance, respect for culture, and conservation of natural resources. The self-evaluation of a community for social usefulness should be considered to complete the circle of acting, observing/measuring, reflecting, improving, and learning (Servaes 2013: 376).

Although development has many dimensions: socio-cultural, political, economic, and environmental, human development should come first. Communication for sustainable social change should be culturally sensitive. Participatory communication and advocacy for sustainable development and the evaluation thereof needs participants who possess what Payutto (1998) calls evolvability (see Chap. 6) and who are interculturally competent.

I was compelled to write this book as a Thai citizen who is concerned about the development of Thailand. Although I spent the last thirty years of my life abroad, Thailand has always been my research interest. So it has for the other contributors to this book.

This book is a compilation of research papers on contemporary Thai issues: crisis in Thai Buddhism, culture, gender violence, tourism, human trafficking, environmental and ecological issues, sustainability and the sufficiency economy, the (mis)handling of elephants, etc.

All four authors have a great interest in Thailand and the development directions it takes. The four authors are specialized and—as we believe in lifelong learning— are specializing in Sociology, Communication for Sustainable Social Change, International Education and Development, and Anthropology and Philosophy respectively.

This book is written from a sociological and anthropological perspective with a dash of communication for sustainable social change. It is comprised of nine chapters which are related and relevant to one another:

Chapter 1 is an explanation of the Thai Buddhist culture from a historical and globalization perspective. Hybridization of Thai Buddhist culture: worldviews or belief systems, values, and representations that reflect Thai patriarchy and nationalism are explicated by Patchanee Malikhao. How Thai Buddhism has become commercialized is explained. How Thai mass media and new media have spurred animism and astrology are further illuminated.

Chapter 2 is the result of online research on the controversial Dhammakaya issue, reported by Patchanee. The Dhammakaya case is a good example of hybridization of the Thai Buddhist culture as a result of the politico-economic influences in the contemporary globalization period. This Chapter is related to the framework of globalization presented in Chap. 1.

Chapter 3 is an assessment of violence towards gender in Thailand from a historical and globalization perspective. This chapter is related to the framework of patriarchy, as presented in Chap. 1. Patchanee would like to draw attention of the reader to sex education as a way to address the problem.

Chapter 4 is an analytical appreciation of Thai culture. Jan Servaes considers Thai culture in essence still a village culture despite of its urbanization and globalization. More importantly, how Thais communicate in relation with power and hierarchy, and how they are influenced by the consumption and production of the media and mass media are being analyzed.

Chapter 5 concerns international relations between Thailand and the People's Republic of China, as Chinese tourists have become the main bread and butter of the tourism industry in Thailand. Tourism and sustainable development is a big issue. Whether Thailand can maintain eco-tourism and conserve its own culture is being discussed by Patchanee.

Chapter 6 touches on globalization, development, and sustainability. The Multiplicity paradigm, which is in line with the late King Bhumibol's Economic Sufficiency paradigm, is discussed. Patchanee shares her experience of working and learning in an organic farm in Chiang Mai, Thailand.

Chapter 7 is one of Patchanee's concerns on the quality of mass communication and journalism in Thailand. Mindfulness is being proposed here as a key to end discrimination and prejudice. How mindful communication and journalism can be achieved is explained from a Thai Buddhist perspective.

Chapter 8 is about human trafficking in Thailand. Fiona Servaes is concerned about the future of poor young people in Thailand, especially young adolescent boys. Everyone knows that young girls and women are vulnerable groups in Thailand. Not many people have paid attention to the plight of young boys. This chapter brings the reader a new perspective on human rights and ways to end the suffering.

Chapter 9 is related to the tourism industry and the mishandling of elephants, the iconic symbol of Thailand. Lisa Servaes gives an anthropological perspective on how the Thai tourism industry and wild animal conservation could find a middle way. Not only human rights but also animal rights need to be considered if one is committed to solutions for sustainable development.

This book may not cover all aspects of modern Thailand, such as Thai pop culture, Thai generation Y, Thai entrepreneurship etc. As influences from other ASEAN cultures can be seen in music, television series, dance performance, movies, and literature, Thai culture has become more infused and hybridized. This may need another book that covers these changes in Thai language, performing arts, relationships, festive celebrations, social media interactions, etc.

I, Patchanee Malikhao, author of six chapters and editor of this book, proudly present this special book to you. Any comments, suggestions, and advice are welcome.

Enjoy reading!

December 19, 2016
Chiangmai, Thailand

References

Martin, J.N., & Nakayama, T.K. (2004). *Intercultural communication in contexts*. Boston: McGraw Hill.
Payutto, P.A. (1998). *Sustainable development* (3rd edition). Bangkok: Buddhatham Foundation.
Sachs, W. (2013). Liberating the world from development. *New Internationalist*. 460, March, p. 25.
Servaes, J. (1999). *Communication for development. one world, many cultures*. Cresskill: Hampton Press.
Servaes, J. (2013). "Communication for sustainable social change is possible but not inevitable". In *Sustainability, participation, and culture in communication* (pp. 369-388). Bristol: Intellect.

Websites

Census.gov, https://www.census.gov/hhes/www/poverty/data/threshold/index.html, accessed March 7th, 2013.

Indexmundi.com, https://www.indexmundi.com/thailand/population_below_poverty_line.html, accessed March 7th, 2013.

The-scientist.com, http://www.the-scientist.com/?articles.view/articleNo/34500/title/Chian-Admits-to-Cancer-Villages-/, accessed March 7th, 2013.

Whereisthailand.info, https://whereisthailand.info/2011/08/saving-accounts/, accessed August 10th, 2013.

Note

Part of this introduction was modified from Malikhao, P. (2013). Communication for Development and Sustainability. *Media Development*, 3, pp. 1–5.

Chapter 1
Thai Buddhism, the Mass Media, and Culture Change in Thailand

Abstract From a historical perspective, Thai Buddhism is a hybridization of animism, Theravada Buddhism, and Brahmanism. As Thailand has gone through four phases of globalization, from the archaic period to proto-globalization, globalization, and contemporary globalization, Thai Buddhist beliefs and practices have also been modified accordingly.This paper attempts to analyze the following:

(1) How the Sangha, or the Buddhist body of Thailand, has been impacted since it has become part of the state during the reign of King Chulalongkorn (Rama V);
(2) How the economic and social development has an impact on Thai Buddhism, especially the animistic beliefs, cults, Hindu Gods, and astrology; and
(3) How the Thai mass media and new social media create hypes on Buddhism, animism, and the further commercialization of Buddhism.

1.1 Introduction

Theravada (ways of the elders) or Hinayana (smaller vehicle) Buddhism has become the main religion of Thailand since the archaic period. According to King Ramkhamhaeng's script, dated 1291, the "Lankawong" doctrine or Theravada Buddhism from Sri Lanka was propagated to the archaic Sukhothai from Nakhon Si Thammarat in the south of current Thailand (Ishii 1986: 60). The soteriology of the Theravada doctrine is "one's acts determine one's salvation" (Ishii 1986: 3). Theravada is the way to cease suffering through the Noble Eightfold Path handed down by the Lord Buddha. Theravadians consider themselves Orthodox Buddhists; they do not mix the Buddhist philosophy with other Eastern traditions and philosophies like the Mahayana (greater vehicle) Buddhism practiced in East Asia (http://viewonbuddhism.org/vehicles.html and http://www.bbc.co.uk/religion/religions/buddhism/subdivisions/mahayana.shtml. Accessed September 22nd, 2014).

As Thailand has passed through many phases of globalization, Buddhism practiced in Thailand has also been affected as part of these cultural changes. Theravada Buddhism has blended with other religious practices such as Hindu,

© Springer Nature Singapore Pte Ltd. 2017
P. Malikhao, *Culture and Communication in Thailand*, Communication, Culture and Change in Asia 3, DOI 10.1007/978-981-10-4125-9_1

notions of power often borrowed from tantric types of Buddhism practiced in Tibet, and animistic folk beliefs in spirits (Baker and Phongpaichit 2005: 9). This sort of hybridization is a result of culture change in many phases of globalization in Thailand.

1.2   Thai Buddhism in the Phases of Globalization

According to Appadurai (2001: 17), globalization is an interactive process in which "locality" and "globality" interact via the shrinking of space-time in the world system. Not only can globality influence locality, but the latter can also induce changes in the global arena and this process is called globalization from below, local globalization or grassroots globalization. Locals get influenced by the culture of globals and become, therefore, "homogenized," Friedman (1994: 210) argues. What happens within the locality is the logical connection process between the decentralization and fragmentation of identities. That creates a new process he refers to as creolization (see also Hannerz 1987). Pieterse (2004) refers to the same process as hybridization. Hawkins (2006: 14) supports this view by stating globalization is multiple and hybrid. However, according to Stuart Hall's work, as studied by Proctor (2004: 27) and Featherstone (1996: 47), globalization concerns both homogenization and hybridization/creolization. As a consequence, each locality is not being hybridized at the same time, speed, or geographical space. This is what Appadurai (1966: 5) refers to as the disparity or disjuncture of globalization, meaning some parts of the world can be more globalized than others. The differences between localities are no longer vertical. Rather, they are horizontal in terms of cultural spaces or nodes connected by crisscrossing flows of people, goods, and messages (Racelis 2006: 55).

Interestingly, globalization is not the same as Westernization. Dicken (2004: 17) states it is not planned; flows happen in many directions and with different degrees, and a globalized locality is not necessarily a Westernized (or Americanized) society. It can lead to some other form of hybridized society. Cohen (1991: 63) states clearly that Thailand did not develop in a Western direction but has fused its own culture with Western forms. Other societies, which have become modern without becoming very Western, are Singapore, Taiwan, and Iran (Nisbett 2003: 224).

According to Hopkins (2002: 1–10), globalization denotes the following ongoing historical process: first, archaic globalization; second, proto-globalization; third, globalization; and fourth, postcolonial globalization. Hopkins explains further that archaic globalization occurred from Byzantium and Tang to the renewed expansionism of Islamic and Christian power after the 1500s. He identified proto-globalization with the political and economic developments that became especially prominent between about 1600 and 1800 in Europe, Asia, and parts of Africa. The third historical process, globalization, he refers to as the colonial period from the 1760s onwards. Globalization that can be related to modernity started from

1800, according to Hopkins. It refers to the rise of the nation state and the spread of industrialization. The last process, postcolonial globalization, refers to the contemporary form that can be dated approximately from the 1950s.

The term "globalization" used in this chapter is defined as a flow of ideas, services, cultural products, and technology that includes the global diffusion and local consumption of culture, values, social, political, and economic concepts. These factors have had an impact, via different communication modes, on a different locality in a different way at a different speed from the archaic past to the present (see also Malikhao 2012).

The periods of globalization used for Thailand as adapted from Hopkins (2002) are (a) *archaic globalization*, starting from the ancient time to the 1500s; (b) *proto-globalization* starting from 1600 to 1768, when Ayutthaya was defeated by Burma for the second time and Siam (former name of Thailand) later shifted the capital city from Ayutthaya to Bangkok; (c) *globalization* starting from 1769 to 1945; and (d) *contemporary globalization* starting from the 1946, when King (Rama IX) ascended to the throne to the present. In each period, hybridization can be observed, and the current Thai Buddhist culture is a consequence of dynamic interplays among the polity, economic, beliefs, worldview, practices, and social change within the globalization and hybridization processes from archaic to present.

Let's start with the religious culture that has changed and hybridized in each globalization period:

In archaic globalization (before 1500s), there were many kingdoms and people had animistic beliefs before Buddhism arrived in the thirteenth century (Rajadhon 1988: 39). The Sukhothai Kingdom can be highlighted as the starting point of the present day Thailand. Sukhothai Kingdom adopted Theravada Buddhism from southward Nakhon Si Thammarat. This adoption process denotes trans-nationalization. Sukhothai Kings had a reciprocal relationship with the Buddha's domain, which has the Sangha (Buddhist monk community) as the center. Sukhothai monarchs entered the monkhood and were supposed to rule with ten Buddhist virtues (Ishii 1986: 61–63). In this period, evidence from inscriptions has shown that the Monarchs conferred the titles of ecclesiastical rank to the Sangha domain. It was a starting point of "state Buddhism" and the hierarchical system in the Sangha (Ishii 1986: 62). Griswold (1967): 13 reports on bronze statues of Hindu gods found in old Sukhothai as good evidence of Brahmanic practices then. Wyatt (1984): 55 stated that although Brahmanism was given court patronage during the Sukhothai period, it did not seem to have any effect on Buddhism at that time. The works of Rajadhon (1986), Keyes (1978), and other scholars in Thai studies also report that animistic beliefs (such as the beliefs in ghosts–phi– and spirit cults) have been side by side with the Brahman rites and Buddhism since this period. The worldview of Thais in this period is based on King Lithai's Trai Phum cosmology from 1345 which consists of three worlds: the world of sensual desire, the world of material factors, and the world of non-material factors (Ganwiboon 2014). http://www.academia.edu/2322014/Cosmology_as_described_in_the_Trai_Phum_Phra_Ruang. Accessed Oct 1st 2014). The first world is about seven realms of happiness: human being, four great kings, 33 deities, full joy, those who are delighted in their full creation, and those who are delighted in creations of others; and four realms of misery: purgatory, animals, suffering ghosts, and the realm of demons. The second world is comprised of 16 realms of Brahma (mind and matters) and the third world is about four stages of immaterial meditation.

The kingdom of Ayutthaya was founded in 1350 and later dominated and engulfed the Sukhothai polity from 1438 onwards (Tambiah 1976: 89). In the early Ayutthaya period, Ayutthaya borrowed Kmer (old Cambodian) civilization through the trans-nationalization process. Tambiah (1976): 89 reports:

> ...the Thai borrowed from the Khmers many features of their administrative and political institutions, art forms, system of writing, and vocabulary, especially that associated with honorific court language. Most importantly, they borrowed the major features of the Khmer royal cult and imported Cambodian brahman priests to conduct the rites.

Although the indigenous Thai village culture is basically matriarchal (Klausner 1997: 64), it shifted to become patriarchal in this period. The impact of patriarchy on sexuality resulted in polygamy, which was recorded for the first time in the early Ayutthaya period. Baker and Phongpaichit (2005: 8) explain that, when one political Tai zone was defeated, the defeated ruler had to send a daughter or sister to become his overlord's wife as a tribute. In special cases, the overlord might grant the subordinate a royal or noble wife.

In the proto-globalization period (from 1600 to 1768), the power of Sukhothai shifted to Ayutthaya. Siam became a large Kingdom. Ayutthaya adopted the devaraja (God-King) cult of polity from Khmer as a starting point of the hierarchical system. This concept is incompatible with Pali Buddhist ideas (Tambiah 1976: 91). This is good evidence that hybridization between Theravada Buddhism and Hindu-Brahmanism occurred.

It was in this period that the *sakdina* system[1] was introduced reinforcing the lower status of females (Baker and Phongpaichit 2005: 16–17). Females in the lower class were subjected to the sakdina lords. It was a tradition that the royals had many wives to ensure the production of enough sons to assist with administrative tasks and enough daughters to build marriage networks within the elite.

Well-established relationships between the Ayutthaya monarchs and the Sangha were recorded in The Chronicles (Ishii 1986: 63). In this period, King Boromakot dispatched a chapter of Buddhist monks to Sri Lanka (Ishii 1986: 64). This is evidence of grassroots globalization in this period. For female subordination, Ghosh (2002: 30–31) explains that, in traditional Siam, the number of wives and female servants indicated the prestige of a man. During this period, spirit-medium cults and magic must have been practiced as shown in a royal decree promulgated by King

---

[1]The sakdina system was the hierarchical structure of service nobility codified in lists of official posts, each with its specific title, honorific, and rank measured in areas of land they were allowed to possess (Baker and Phongpaichit 2005: 15, Ongsakul 2005: 115, Servaes and Malikhao 1989: 33, Servaes 1999: 211, Srisootarapan, 1976; Suwannarit 2003: 9–12).

Rama I in 1782. The then King warned the Thai subjects not to pay much attention to spirits and magical things while ignoring the Buddhist teaching (Kitiarsa 1999: 94).

In the globalization period (1769–1950), Siam (former name of Thailand) entered modernity. The power of Ayutthaya shifted to Dhonburi and later to Bangkok. King Rama I of *Rattanakosin* (the starting of the Chakri Dynasty with Bangkok as the capital city) era ordered all monks to have an ID. Bad monks were defrocked. Christian missionaries came in the late 1820s and that made Monk-Prince Mongkut reform the Sangha by establishing the Thammayut movement in the reign of King Rama III (Ishii 1986: 65 and http://guru.google.co.th/guru/thread?tid=59242e591b555f35. Accessed September 23rd, 2014). The Thammayut Nikaya is distinctively different from the Lankawong or Maha Nikaya (mainstream Lankawong Theravadin monks) in emphasizing stricter vinaya (disciplines) and focusing on meditation (http://guru.google.co.th/guru/thread?tid=59242e591b555f35. Accessed September 23rd, 2014).

King Rama IV, also known as King Mongkut (reigned from 1824–1868), reformed Buddhism in terms of the interpretation of Dhamma (the truth or nature of the world as described by the Lord Buddha) by editing ancient Buddhist texts and propagating a reformed version to suit the modernization era based on Science (Visalo 2003: 8–11). Meditation practices and metaphysical miracles were eradicated from the Buddhist studies curricula for monks. Heaven and hell were explained as a state of mind, rather than as the world out there after death. The worldviews of the people had changed to greater secularism and the image of Buddha and Kingship was reduced from divinity to a more human form (Visalo 2003: 18–30).

During the reign of King Rama V, also known as King Chulalongkorn (1873–1910), Thailand entered the modernization period. Out of fear of colonization by Great Britain and France, King Chulalongkorn initiated fiscal, education, communication, and transportation reforms with Bangkok as the center (Charoensin-o-larn 1988: 139 and Sivaraksa 2001: 33–34). Although Winks (1976) suggests that the concept "informal empire" can be used for such a reform, Holm (1997: 124) explains that the informal form of imperialism imposed on Thailand revolved mostly around the control of railway construction from Bangkok, and the degree of involvement was limited to the cooperation of the royal family and the upper aristocracy. A formal form of imperialism was imposed on Siam when it signed the "Bowring Treaty" with the British in 1851. It aimed to reduce import duties and taxation and allowed British subjects to reside in Siam and enjoy rights of extraterritoriality from the Siamese courts (Wyatt 1984: 184). Siam also signed Bowring-like treaties with the USA, France, and a score of other states (Wyatt 1984: 184). Charoensin-o-larn (1988): 139 calls the way King Rama V transformed the traditional decentralized sakdina (Thai feudal) states into a highly centralized and unified state under absolutism as "internal colonization." I would rather call the way Kings Rama V, VI, and VII rearranged the structure of the country to meet Western standards in the globalization period as a *hybridization of the Thai political-economy.*

King Rama V, with the assistance of his half-brother, the Monk-Prince Vajirananavarorasa, modernized the ecclesiastical education (Payutto 2001: 137). Payutto further reports that the Royal Siamese Tripitaka or the first complete set of the Pali Canon was published; a royal library was founded to preserve Buddhist sacred books and rare scriptures. However, Buddhist monks in Siam lost their important position in education when King Rama VI reformed national education from temple-based to school-based (Klausner 1993: 160; Payutto 2001: 137). Payutto (2001: 140–141) stated that Buddhist monks were considered part of the traditional Thai society. As the Thai education system aims at Westernization, Buddhist monks confined their activities to merit-making acts, preaching the Precept Five on basic morality, focusing on monastic affairs, and construction of monastic buildings. Some monks engaged in superstition and astrology. Payutto (2001: 143) observes:

> Modern Thailand is, however, often branded with modernization without development or with misguided development. The lack of the monks' share in the process of development must have been a factor in this undesirable result.

In contrary to the modernity asserted by the elites in Siam, Crawfurd (1967), a British diplomat to Siam, reported the intermingled practices and beliefs of the Siamese and the Indian and Chinese immigrants in 1822. Crawfurd observed the worship of linga (phallic symbol) of Brahmanism, Hindu Gods and Goddesses, and Chinese deity, such as Kuan Yin. Alongside with Brahmanism, animism, and supernaturalism were being practiced as reported by Kitiarsa (1999:77–82).

In the contemporary globalization period (1946-present), Thailand entered postmodernity. From 1970 s onwards Thailand has been under the flux of contemporary cultural globalization via modern telecommunication technology, transportation, mass media, and the Internet. Vuttanont et al. (2006: 2069) explain Thailand in transition from:

> (1) feudal towards neo-capitalist political system[2]; (2) from restricted towards widespread information; (3) Buddhism towards multi-faith or secular; (4) from high towards low religiosity; (5) from rural towards urbanized geography; and (6) from the following social values to the new ones: from respect the old towards celebrates youth; from collectivism towards individualism; from trusting towards sceptical; from modesty towards self-expression; and from male dominated towards gender equality.

Bechstedt (2002: 238–261) explains that the hierarchical system of the past mixed with the emergence of new institutions formed by interest groups and new social classes. Those in power have money and access to profits, shares, and stocks. Khun Ying Amporn Meesuk, a renowned Thai scholar, interviewed by Trisuriyadharma (2006: 17) suggests that now the whole Thai society worships money as God. Thai society is now facing a paradox between maintaining its traditional culture and adjusting to new changes. Techapira (2006), a well-known

---

[2]Teharanian (2007: 91) explains neo-capitalism as the incorporation of capital and global reach of transnational corporations (TNCs), dominating state, civil societies, and communication networks; disembodiment of human relations into a nexus of digital numbers.

Thai social critic, states Thai society has suffered cultural schizophrenia as Thais are trapped in a double bind between Westernization and their own traditional culture. According to him, being traditional Thai means being authoritarian, being pragmatic, and being subservient as part of a patronage system. These characteristics are opposite to being egalitarian, self-reliant, and ideological according to universal principles. Techapira calls the hybridized way of Thais adjusting themselves to universal principles but still maintaining some Thai characteristics a relativistic movement.

Baker and Phongpaichit (2005: 150–164) write that in this period, the USA mainly supported economic development, educational development, and bureaucratic infrastructure building for the promotion of development. The USA also provided funding to fight communism during the cold war period. Thailand could be seen as part of an informal US empire in this period and accepted the Social and Economic Development Plan from the Free World Leaders (Thongchai 2001: 37). As a result of economic development, Thai–Chinese entrepreneurial groups were in control of the economy. It is fair to say that Thailand in this globalization period follows the Modernization paradigm, which sees the industrialized Western societies as the ultimate goal of development. As explained earlier, Thais worship spirits as part of their animistic beliefs since the archaic globalization. Urbanization and migration as part of Westernization have caused the booming of the urban-based spirit-medium cults in the past few decades as a moral and psychological refuge for the capitalistic desire of the urban population (Kitiarsa 2012: 16–17, 146). Two noteworthy points are (1) the rural spirit mediums are normally female who assert themselves from the subordination in the Thai patriarchal system to possess power above laymen but lower than Buddhist monks, and (2) the urban spirit cults incorporate the Chinese and Hindu cults, not the American nor the European ones due to the inaccessibility of less educated rural spirit mediums and the surveillance of the Sangha to limit the hybridization form to the trio of Buddhism–Hinduism–animism (Kitiarsa 2012: 52–54, 147). Mahamakuta and Mahachulalongkorn Rajawittayalai were established as Buddhist universities in 1946 and 1947, the teaching and learning emphasized rote learning of Buddha-Dhamma (teachings of Buddha in Pali).

Consumerism associated with hybridized forms of Buddhism, supernaturalism, and animism can be seen in the growth of the expensive amulets and talisman industry, which involved famous magic monks (see Formoso 2000: 99, Suntravanich 2005 quoted in Prachachart 2003: 13). Magic monks, according to Kitiarsa's study (Kitiarsa 2012: 39–40), are those who perform all or some of the following: eliminating of bad omens/strengthen good fate (sado khro); fortune-telling (du duang); spraying or bathing a person with blessed water to ward off bad luck and protect the person from bad spirits (rot nam mon); enhancing the longevity and well-being (to ayu/suep chata); expelling or exorcising bad spirits (lai phi/khap pop); setting up a guardian spirit's house or altar (yok san phra phum); blessing a new car or new properties (choem rot/ban/sammak ngan); and providing tips for lotto numbers (bai huay/hai chok hai lap). Pluralism of popular Buddhism is attested to by the emergence of more than 100 cults and sects of animism–

Buddhism (Visalo 2003: 176). Kitiarsa studied the religious practices in the contemporary modernization period and collected popular spirit-medium cults (2012: 23–30); they are Buddha as the supreme deity; spirits of famous Buddhist saints or magic monks, Indian gods and goddesses, Chinese deities, royal spirits (spirits of late great Kings and Queens), and local guardian and tutelary spirits. This phenomenon coins with what Payutto (2001): 153 sees as extreme modernization that causes secularization or even politicization. Kitiarsa (2012): 118 also confirms that amulets exemplify the growing worldliness of Thai religiosity and a junction of religious practice and political anxieties. The politicization of Buddhism can be seen from the involvement of the Santi-Asoke with the Palang Dhamma political party (*Matichon Daily,* February 27, 2006) and the Dhammakaya with Phue Thai political party. The Thai-Chinese middle class was drawn to these new religious movements: Buddhadasa Bhikku, Santi-Asoke, and Dhammakaya. The first one is an intellectual and self-taught Buddhist monk who focused on a vanguard interpretation of Buddhism, the second on a skewed interpretation of Buddha's teaching (later the abbot was defrocked), and the last one on commodification of Buddhism as well as a skewed interpretation of Buddha's teaching to serve material wealth (see Payutto 1988, 2008; Formoso 2000: 101; Ekachai 2001: 93–104). Buddhadasa's work is in line with the ecumenical movement of socially engaged Buddhism that concerns both Mahayana and Theravada Buddhism (Schober 2012: 16). These later two movements attained some popularity because popular Buddhism had become tainted with sex, money, title scandals, and astrology; and also because of the reluctance of the Sangha to reform itself to solve its so-called spiritual ailings due to consumerism and the materialism associated with capitalism as well as its reluctance to allow females to be ordained as monks (Ekachai 2001).

The mixture of Theravada Buddhism with Brahmanism, supernaturalism, and belief in spirits (animism) has played an important role in forming the Thai belief and values system as I will discuss in more detail next.

1.3　Thai Culture: A Mediatized Buddhist Culture

Klausner (1997), Mulder (1985, 1990) and Servaes (1999) studied Thai culture and described the culture as a village culture with mutual trust and informal social relationships among the inner-groups, but with distrust and formality-oriented relations toward the outer-groups. The distrust of strangers leads to the belief in supernatural power to protect themselves from dangers or the seeking of protection from wealthy or powerful people. The latter was known as the entourage concept or the patronage system in Thai society (Komin 1990: 155).

Servaes (1999: 12) suggests that a culture can be analyzed by its four distinguishable but interrelated analytical components: a worldview (*Weltanschauung*), a value system, a system of symbolic representations, and a social organizational system. For Thai culture, the worldview of Thais is comprised of patriarchy and a unique Buddhist worldview, which incorporates animistic, supernatural beliefs, and

Brahmanism (Malikhao 2007: 63). Srichampa affirms that Thais have different beliefs (2014: 50–51): Buddhist beliefs, superstitious beliefs, sacred thing beliefs, deity beliefs, and astrological beliefs. For Buddhist beliefs, Srichampa (2014: 50) explains that these are about the triple gems (Buddha-Dhamma-Sangha), karma law (do good get good, do bad get bad), reincarnation, law of nature which consists of the Law of Kamma (action), the Law of Season (utu), the Law of Seed (Bija), the Law of Consciousness (Citta), the Law of Dhamma (States), and heaven and hell. For superstitious beliefs, Srichampa (2014: 51) explains that there are two types: magic beliefs—the beliefs of old scripts—and amulet beliefs. These can be called the popular Thai Buddhist beliefs. Even the current Prime Minister, Gen. Prayudh Chan-Ocha, revealed to the reporters on September 17, 2014, that he carries an elephant hair bangle, Buddha ring made out of 9 gemstones (for good luck and prosperity), and a ring he received from the Queen (as the Queen's musketeer).

Mediums, amulets, magic monks, spirit-of-ancient-royal cults, ghosts-of-superstar cults mentioned earlier are symbolic representations of these beliefs. Stenges (2009: 4) studied King Chulalongkorn cult where people worship the spirit of the King and ask for his advice through mediums and concluded that mediation between modernity and Thainess is the central theme in the King Chulaongkorn cult. Stenges (2009: 24) explains that "the King Chulalongkorn cult is but one among many "junctions" that make up Thai public culture today. Since the 1960s and particular in the 1980s, the Thai religious realm has been flooded by a wave of cults around (historical kings, queens, monks, local heroes and heroines, gods and goddesses)".

The core value of Thais revolves around the core ideology of nation-king-religion adopted from England in the reign of King Rama VI or King Vajiravudh (Cohen 1991: 15).

Worldview is also a part of the construction of culture. It is directly influenced by long-established traditional beliefs and religion. Berger and Luckmann discuss in "The Social Construction of Reality" (1966) the interaction between thinking and action. Socialization within a tradition and culture shapes an individual's thinking, and at the same time, this internalized form is reflected in the manifestation of culture (Holm 1997: 75). This model stresses the importance of religion, as it provides a symbolic universe that explains birth, life, and death, as well as providing the individual with an identity. Religion explains the world through myths and legends and also through rational discourses. Therefore, Robertson (1972: 47) defines *religious culture* as: "… a set of beliefs and symbols (and values deriving there from) pertaining to a distinction between an empirical and a super-empirical, transcendent reality; the affair of the empirical being subordinated in significance to the non-empirical."

Smart (1983: 7–8) presents six dimensions of religion: the doctrinal, the mythic, the ethical, the ritual, the experiential, and the social. A religion typically has a written system of doctrines; this is the doctrinal dimension. It has a special story with a sacred meaning to pass on to believers; this is the mythic dimension. It sets certain rules of do's and don'ts and precepts; this is the ethical dimension. It involves its believers in religious action such as worshipping, praying, singing hymns, listening to sermons; this is the ritual dimension. McGuire points out many symbolic representations of religion in terms of discourses that we can observe and analyze, such as rituals, symbols, religious experiences (McGuire 2002: 124–125).

The mass media audience tends to follow the notion which Festinger (1957) proposed about cognitive dissonance, that people tend to avoid adopting messages and information that are not congruent to their existing worldview. Hence, people are looking for a confirmation of their bias, rather than for genuine information. Sets of cultural products shared among many localities are what constitute "popular culture" (which was in the past labeled as "low culture"—as an opposite to "high culture" shared by the elites such as classical music, opera, and ballets). The concept of popular culture has been discussed by scholars such as Burke, Evans-Prichard, and Geertz (http://www.answers.com/topic/popular-culture. Accessed January 25, 2012). Nachbar et al. (1978: 6–8) explain that examples of popular mythologies are beliefs, values, superstitions, and actual myths; popular artifacts are, for example, product packaging, architecture, toys and icons, and logos; popular arts and performing arts such as rock and roll, and films; and popular rituals such as the Olympics, concerts, holidays, and festivals. Therefore, Holmberg summarizes popular culture as follows:

> ... popular culture includes the human activities, languages, and artefacts that grow and nourish people in communities and that generate observable, describable interest about its events and artefacts, within a community and between communities (Holmberg 1998: 15).

The production and distribution of communication controlled by the communication industry promotes popular culture. The industry, according to Macbride et al. (1980: 96–97), consists of printed media enterprises, radio and television companies, news and features agencies, advertising and public relations firms, syndicates and independent companies producing and distributing print, visual and recorded material for print and broadcasting conglomerates, public or private information offices, data banks, software production, manufactures of technological equipment, and so on. Productions from the communication industry are also known as the cultural industry because they record and reproduce cornucopia of social interactions, representations, and organization systems in diverse media forms such as books, arts, films, recordings, television, radio, the internet, plays, concerts, and music. With the breakthrough of the new media as a consequence of the digitization revolution, the symbolic representations of popular culture rapidly transmitted by the information super highway create diverse interpretations of self and identity, sex, gender, sexuality, religious practices, beliefs, etc. The emergence of modern Buddhist institutions and social roles, worldviews, ethics was a reflection and inflection of the politico-socio-economic ideologies (Schober 2012: 16). I would like to contribute this effect to the framing and rituals of the communication industry. With the help of new media such as the internet and satellite TV, many new religious movements target their audience to world peace and social harmony (Schober 2012: 22, Agarwal 2014: 357–359).

Without the mass media as weaving threads in the globalization process, the beliefs, religious practices, values, and cults would not have had a great impact.

Kitiarsa (2012: 66, 67, 92, 102, 118, 126, 143) studied three examples of cults around the spirit of Phumphuang Duangchan, the late country music queen, and the lotto mania; Luang Pho Khun, the famous magic monk and his amulets that bless people to become rich and free from dangers; and Chatukham-Ramathep, an invented Thep or divine spirit for commercialization purposes and concluded that the mass media had constructed and reconstructed the stories around these three personas and created hyperreality. The media cited are Thairath, Matichon, Khao Sot newspapers, talkshows on TV such as the Cho Chai (pierce through the heart) program, Lok Thip (Divine World) and Mahalap (great fortunes), and Daraphapayon (movie stars) magazines. According to Kitiarsa (2012), the media framed the personal stories of these three examples to the interests of the public in spirit-medium cult and amulet cult. Couldry (2003: 26) discusses "media rituals" which he explains as a symbolic expression of any social relations by the media. The media created "myths" around the persona in question. In Phumphuang's case, the media enhance her tragic death through framing with supernaturalism; in Luang Pho Khun's case with his charismatic characters and magic amulets that induce good fortune, wealth and safety; in Chatukham-Ramathep with invented divinity that blesses good luck, fortune, and wealth. The rituals of people giving gifts and worshipping the effigy of Phumpuang became the media ritual as well as people go for the holy knocking on the head and holy spray from Luang Pho Khun or the mythification/inauguration process of Chatukham-Ramathep amulets. Recently in 2016, the media created a hype of Tukata Look Thep or Divine Dolls. Thai newspapers spread the news that airline companies allow the believers of Tukata Look Thep to book a seat on a plane for their dolls (Dailynews online, February 29, 2016). Some believers carry around a doll and treat it as if it was their child out of the belief that the doll would bring them good luck. The dolls were found of containing hair or a piece of nails from a dead body and were believed to have spirits inside. These were stories portrayed by the mass media.

Kitiarsa (2012: 84) refers to the process by which the mass media enhance and create hypes around medium-spirit cults "mediation." According to him, mediation goes hand in hand with deification and commodification (Kitiarsa 2012: 84). I would rather use the term "mediatization" as it is the process whereby culture and society are increasingly dependent on the media and their logic in such a way that the degree of the social interactions within a given culture and society modulated by the media capital can be observed within social institutions, between institutions, or in a society (Hjarvard 2013:17). It is seen as a longer term process than mediation as the media's influence on the change of the social and institutional interactions is to be expected whereas the mediation describes only the concrete act of communication by using a type of media in a given social context (Hjarvard 2013: 18–19).

Mediatization, as a part of globalization, characterizes the postmodern culture industry. The productions and products of the communication and cultural industry record and reproduce cornucopia of social interactions, representations, and organization systems in diverse media formats. Mediatization is closely related to individualization as I will explain next.

Individualization can be traced back to the study of Carl Jung's work on individuation (http://www.mindstructures.com/carl-jung-individuation-process/. Access March 27, 2014). He stated that

it is a process by which individual beings are formed and differentiated; in particular it is the development of the psychological individual as a being distinct from the general, collective psychology.

According to Jung, the ego is a subject of one's consciousness and a subset of the self, which includes both the consciousness and unconsciousness of the total psyche. Globalization increases this individualization process. Consequently, the German Sociologist Ulrich Beck proposed his individualization theory, arguing how individuals in the late modernity take their lives in their own hand (due to migration and economic opportunities) and thereby create a new identity of social life (http://damienlanfrey.net/web/index.php?option=com_content&view=article&id=12:notes-on-ulrich-becks-qindividualizationq&catid=13:modernity. Accessed March 27, 2014).

With the breakthrough of the new media as a consequence of the digitization revolution, new formats of self-expressions have become popular as the audience gains recognition of his/her private, social, and public achievements (Hjarvard 2013: 150). Hjarvard (2013:11) explains that the mediatization process affects an individual autonomy and social belonging in such a way that the individual gains more autonomy by relying deeply to the external world in the act of connecting to the large social networks available. He calls this phenomenon, *soft individualism*. Elliott and Lemert already observed a new kind of individualism in 2006. They propose that globalization has a profound impact on the individual level. They defined this *new individualism* as a highly risk-taking, experimenting, and self-expressing individual underpinned by new forms of apprehension, anguish, and anxiety. High levels of individualism can lead to *narcissism*. Twenge and Campbell (2009: 19) state in their book, "The Narcissism Epidemic," that the central feature of narcissism is a very positive and inflated view of self and this value is growing rapidly in the American culture fueled by the mass media, including the new media, and changes in parental approaches to upbringing that emphasizes self-expression. Symbolic representations of the new American culture of self-expression or participating audience/amateur journalist are the emphasis on celebrities in the media, the success of Facebook as social networking sites, the uploading of personal videos on YouTube, twitter (micro blogging and text-based social networking or SMS on the internet via its own website), and blogging (Twenge and Campbell 2009). In many cases the audience can be a target or a commodity when profit-making and commercial values are built into the media system of advertising. The self or ego of the audience will in these cases be coupled with commercial products to increase self-confidence, self-respect, self-esteem, etc.

The mediatization effect can be used to explain the booming of spirit-medium cults, amulet cults, and commercialized religious movements as external factors to help strengthen the new kind of individualism.

1.4 Conclusion

Trends of Theravada Buddhism in Thailand conform to the global trend that we see the rise in secularization but it is still within the realm of the trio of Theravada Buddhism–Hindu–Brahmanism–Animism. The hybridized Thai popular Buddhism is a consequence of the interplay of politico-socio culture from a historical perspective with the free market economy, advertisement, the mass media, and charismatic figures. Without a doubt, the mass media construct myths and reconstruct the personages to suit the framing of popular religious practices in the so-called mediatization process. The booming of talisman and amulet industry as well as new cults and religion movements as part of religious culture change is the answer to the new individualism in the postmodernity. Hence, this paper has answered the main research questions:

(1) The Sangha,or the Buddhist body of Thailand, has been impacted since it has become part of the state during the reign of King Chulalongkorn (Rama V) through reformation of education and centralization.
(2) The economic and social development in the contemporary globalization process has created monetary needs and anxiety to compete in a postmodernity and the Thai popular Buddhism, especially the animistic beliefs, cults, Hindu Gods, and astrology, emerged to respond to the new social needs and new consumerism values.
(3) Thai mass media and new social media create hypes on Buddhism, animism, and the further commercialization of Buddhism via framing and media ritualization of icons to suit the needs for magic, refuge, blessing, good luck, and prosperity as the individual has changed into a new kind of individualism as a result of mediatization.

References

Agarwal, R. (2014). Marketing and branding of religions in Thailand: The importance of social media. *The 7th international forum on public relations and advertising proceedings August 13th–15th* (pp. 355–361). Bangkok: Mahidol International College.

Appadurai, A. (1996). *Modernity at large: Cultural dimensions of globalization*. Minneapolis and London: University of Minneapolis Press.

Appadurai, A. (2001). Grassroots globalization and the research immagination. In A. Appadurai (Ed.), *Globalization* (pp. 1–21). London: Duke University Press.

Baker, C., & Phongpaichit, P. (2005). *A history of Thailand*. New York: Cambridge University Press.

Bechstedt, H. (2002). Identity and Authority in Thailand. In C. J. Reynolds (Ed.), *National identity and its defenders: Thailand today*. Chiangmai: Silkworm Book.

Charoensin-o-larn, C. (1988). *Understanding postwar reformist in Thailand*. Bangkok: Editions Duangkamol.

Cohen, E. (1991). Christianity and Buddhism in Thailand: The Battle of the axes and the contest of power. *Social Compass, 38*(2), 115–140.

Couldry, N. (2003). *Media rituals. A critical approach*. London and New York: Routledge.
Crawfurd, J. (1967). *Journal of and embassy to the Court of Siam and Cochin China*. Kuala Lumpur: Oxford University Press.
Dicken, P. (2004). Globalization, production and the (im)morality of uneven development. In R. Lee. & D. M. Smith (Eds.), *Geographies and moralities: International perspectives on development, justice, and place* (pp. 17–31). Malden, MA and Oxford, UK: Blackwell Publishing Ltd.
Ekachai, S. (2001). *Keeping the faith: Thai Buddhism at the crossroads post books*. Bangkok: The Post Publishing Plc.
Featherstone, M. (1996). Localism, globalism, and cultural identity. In R. Wilson & W. Dissanayake (Eds.), *Global local: cultural productions and the transnational imaginary* (pp. 46–77). Durham and London: Duke University Press.
Festinger, L. (1957). *A Theory of cognitive dissonance*. Stanford, CA: Stanford University Press.
Formoso, B. (2000). *Thaïlande: Buouddhisme Renoncant Capitalisme Triomphant*. Paris: La Documentation Francaise.
Friedman, J. (1994). *Cutural identity & global process*. London: Sage Publications Inc.
Ganwiboon, S. (2014). Cosmology as Described in Trai Phum Phra Ruang http://www.academia. edu/2322014/Cosmology_as_described_in_the_Trai_Phum_Phra_Ruang. Accessed October 1, 2014.
Ghosh, L. (2002). *Prostitution in Thailand: Myth and reality*. New Delhi: Munshiram Manoharlal Publishers Pvt. Ltd.
Griswold, A. B. (1967). *Towards a history of Sukhodaya Art*. Bangkok: Fine Arts Department.
Hannerz, U. (1987). The world in Creolization Africa. *Journal of the International African Institute, 57*(4), 546–559.
Hjarvard, S. (2013). *The mediatization of culture and society*. London and New York: Routledge.
Hawkins, M. (2006). *Global structures, local cultures*. Melbourne: Oxford University Press.
Holm, N. (1997). An Integrated role theory for the psychology of religion: Concepts and perspectives. In B. Spilka & D. McIntosh (Eds.), *The psychology of religion* (pp. 73–94). Oxford: Westview Press.
Holmberg, C. B. (1998). *Sexualities and popular culture*. Thousand Oaks: Sage Publications.
Hopkins, A. G. (2002). Introduction: Globalization—an agenda for historians. In A. G. Hopkins (Ed.), *Globalization in world history* (pp. 1–10). London: Pimlico.
Ishii, Y. (1986). *Sangha, state, and society: Thai Buddhism in history* (P. Hawkes, Trans.) Monographs of the Center for Southeast Asian Studies Kyoto University. Honolulu: The University of Hawaii Press.
Keyes, C. F. (1989). *Thailand, Buddhist kingdom as modern nation-state* (1st ed.). Bangkok: Duang Kamol.
Kitiarsa, P. (1999). You may not believe, but never offend the spirits: Spirit-medium cult discourses and the postmodernization of Thai religion. Doctoral dissertation. University of Washington.
Kitiarsa, P. (2012). *Mediums, monks, & amulets. Thai popular Buddhism today*. Chiang Mai: Silkworm Books.
Klausner, W. (1993). Popular Buddhism in North East Thailand. In W. Klausner (Ed.), *Reflections on Thai culture* (pp. 159–176). Bangkok: The Siam Society under Royal Patronage.
Klausner, W. (1997). *Thai culture in transition*. Bangkok: The Siam Society.
Malikhao, P. (2007). HIV/AIDS strategies in two Thai communities: Buddhist and Christian. Ph. D. thesis: the University of Queensland.
Komin, S. (1990). *Psychology of the Thai People*. Bangkok: National Institute of Development Administration (NIDA).
Malikhao, P. (2012). Sex in the village. culture, religion and HIV/AIDS in Thailand. Penang/ChiangMai: Southbound/Silkworm Books.
McGuire, M. (2002). *Religion: The social context*. Belmont: Wadsworth Thomson Learning.
Mulder, N. (1985). *Everyday life in Thailand: An interpretation*. Bangkok: DK Books.
Mulder, N. (1990). *Inside Thai society: An interpretation of everyday life*. Bangkok: DK Books.

Nachbar, J., Weiser, D., & Wright, J. L. (1978). *The popular culture reader*. Bowling Green: Bowling Green University Popular Press.

Pieterse, J. N. (2004). *Globalization and culture*. Oxford: Rowman & Littlefield Publishers Inc.

Prachachart, S. (2003, 21st–27th October). 10,000-Million Amulet Business: The World Most Expensive Mass. *Matichon Weekly, 23*, 13.

Nisbett, R. E. (2003). *The geography of thought: How Asians and Westerners Think Differently and Why*. New York: The Free Press.

Ongsakul, S. (2005). *History of Lan Na (C. Tanratanakul, Trans.)* (1st ed.). Chiang Mai: Silkworm Books.

Payutto, P.A. (2001). Buddhism in contemporary Thailand. In *Thai Buddhism in the Buddhist World. A survey of the Buddhist Situation against a historical background.* (pp. 137–154). Bangkok: Buddhadhamma Foundation.

Proctor, J. (2004). *Stuart hall*. London: Routledge Taylor & Frances Group.

Racelis, M. (2006). Globalization and people: Human agency in a wired world. In S. Thipakorn & S. Wun'gaeo (Eds.), *Globalizing economy and civilizational agenda* (pp. 43–57). Bangkok: Institute of Asian Studies, Chulalongkorn University.

Rajdhon, P. A. (1988). *Essays on Thai folklore* (3rd ed.). Bangkok: Thai Inter-religious commission for development.

Robertson, R. (1972). *The Sociological Interpretation of Religion*. Oxford: Basil Blackwell.

Schober, J. (2012). Modern Buddhist conjunctures in Southeast Asia. In D.L.McMahan (ed.), *Buddhism in the modern world*. London and New York: Routledge.

Servaes, J. (1999). *Communication for development: One world, multiple cultures* (1st ed.). Cresskill, New Jersey: Hampton Press Inc.

Servaes, J., & Malikhao, P. (1989). How 'culture' affects films and videos in Thailand. *Media Development, 36*(4), 32–36.

Sivaraksa, S. (2001). *Manodharmsamneuk samrab Sangkom Ruamsamai (Conscience for Contemporary Society)*. Bangkok: Religion for development committee.

Smart, N. (1983). *Worldviews: Crosscultural Explorations of Human Beliefs*. New York: Charles Scribner's Sons.

Srichampa, S. (2014). Thai Amulets: symbol of the practice of multi-faiths and cultures. In P. Liamputtong (ed.), *Contemporary Socio-Cultural and Political Perspectives in Thailand.* (pp. 49–64). London: Springer.

Srisootarapan, S. (1976). *Chom Nar Sakdina Thai (The Face of Thai Sakdina)*. Bangkok: Agsorn Sampan.

Stenges, I. (2009). *Worshipping the great Modernizer King Chulalongkorn, Patron saint of the Thai Middle Class*. Singapore: NUS Press.

Suwannarit, V. (2003). *Withi Thai (Thai ways)*. Bangkok: Odian Store.

Tambiah, S. J. (1976). World conqueror & world renouncer. A Study of Buddhism and polity in Thailand against a historical background. Cambridge: Cambridge University Press.

Techapira, K. (2006). *Critiques of Thainess: Critiques of critiques* (p. 6). Thai: Matichon Daily.

Thongchai, P. (2001). Patirupe Karn Sueksa: Pue Kwam Khaeng Kraeng Khong Pollameung Thai (Education Reform for the Strenght of Thai Citizens). In S. Sophonsiri (Ed.), *Withi Sangkhom Thai (Thai Ways): Sarnniphon Tang Wichakarn Naueng Nai Wara Neung Sattawat Pridi Bhanomyong Chude Tee Jed (Academic monograph series 7 in the occasion of the 100th anniversary of Pridi Bhanomyong) Karn Sueksa Pua Kwam Pen Thai (Education for Freedom)* (pp. 26–103). Bangkok: Pridi Bhanomyong Foundation.

Trisuriyadharma, P. (2006). *Interview of Khun Ying Amporn Meesuk, President volunteers for the social foundation adn committee of national human rights*. Bangkok: Kled Thai.

Twenge, J. M. & Campbell, W.K. (2009). Living in the Age of Entitlement. The Narcissism Epidemic. New York: Free Press.

Visalo, P. (2003). *Thai Buddhism in the future: Trend and ways out of crisis (in Thai language)*. Bangkok: Sodsri-Saritwong Foundation.

Vuttanont, U., Greenhalgh, T., Griffin, M., & Boynton, P. (2006). Smart boys and sweet girls–sex education needs in Thai teenagers: A mixed-method study. *The Lancet, 368*(9552), 2068–2079.
Wyatt, D. K. (1984). *Thailand: A short history*. New Haven and London: Yale University Press.
Winks, R. W. (1976). On Decolonization and Informal Empire. *The American Historical Review, 81*(3), 540–556.

Websites

http://www.answers.com/topic/popular-culture. Accessed Jan 25, 2012.
http://damienlanfrey.net/web/index.php?option=com_content&view=article&id=12:notes-on-ulrich-becks-qindividualizationq&catid=13:modernity. Accessed March 27th, 2014.
http://guru.google.co.th/guru/thread?tid=59242e591b555f35. Accessed September 23rd, 2014.
http://guru.google.co.th/guru/thread?tid=59242e591b555f35. Accessed September 23rd, 2014.
http://www.mindstructures.com/carl-jung-individuation-process/. Accessed March 27th, 2014.
http://www.dailynews.co.th/economic/375084. Accessed February 29th, 2016.

Note

This article is a reprint of Malikhao, P. (2015). Thai Buddhism, the Mass Media and Culture Change in Thailand. In *Journal of the Asian Research Center for Religion Communication*. 12 (2), pp. 124–143.

Chapter 2
Analyzing the "Dhammakaya Case" Online

Abstract In this chapter, the hybridization of Thai Buddhism in the contemporary globalization period will be assessed through the so-called *Dhammakaya case*. Dhammakaya will be analyzed as one of the controversial approaches in relation to Hinnayana Buddhist doctrines, using the opinion of two experts on Hinnayana Buddhism—Buddhadasa and P.A. Payutto—as resources. An interview of another intellectual monk, Phra Paisal Visalo, has also been included. This chapter attempts to assess the information discussed on digital social media such as Twitter, Internet, and Instagram related to the Dhammakaya Temple and its abbot. Content and contextual analyses will be conducted. The research findings will be discussed against the background of modernization, Thai Buddhist culture, and digital communication in Thailand.

2.1 Introduction

During the globalization period in Thailand, Buddhist temples, which used to be the hub of moral ethic and academic education, have lost their grip on the Thai educational system, as the Minister of Education took over the monks' role in formal education in the reign of King Rama V or King Chulalongkorn (Educational Management Information System Centre 1998). Buddhist monks have since confined their roles to spiritual guide, and kept themselves busy on Buddhist or Pali studies, social work related to the further amelioration of temples, and ritual services such as giving sermons and praying. (Payutto 2001: 153).

During the contemporary globalization period, Thai Theravada Buddhism has become under scrutiny from the public. Some monks provide services that Buddha did not endorse, such as fortune-telling, making sacred amulets, and giving lotto numbers. (Kitiarsa 2012: 52–54).

In this period, many scandals related to monks have been recorded (Ekachai 2001). Sexual laxity by monks, while professing Buddhism, is illustrated by the following cases: Phra Yantra Amaro Bhikku who fled to the USA after an accusation of his fathering a child with a female follower in 1994; Phra Nikorn

© Springer Nature Singapore Pte Ltd. 2017
P. Malikhao, *Culture and Communication in Thailand*, Communication, Culture and Change in Asia 3, DOI 10.1007/978-981-10-4125-9_2

Dhammawatee who was defrocked in 1990 on the grounds of making a female follower pregnant; and Phra Pawana Buddho who was arrested for raping underaged hill tribe girls studying under his scholarship program in 1995 (Ekachai 2001: 60, 111). Nane (Novice) Kham and his donation embezzlement news was a big issue in Thai society in 2013–2014, before he fled to live in the USA (Manager online 2013). Recently, news about Wat Phra Dhammakaya has been on the front pages of Thai newspapers. It attracted the attention and action from the Department of Special Investigation (DSI). The ongoing online discussion on social networks and in some mainstream media has been intense.

Consumerism associated with the hybridized form of Buddhism and animism can be seen in the growth of the expensive amulets and talisman industry, which involved famous monks (see Formoso 2000: 99, Suntravanich 2005 quoted in Prachachart 2003: 13). Pluralism of popular Buddhism is exemplified by the emergence of more than 100 cults and sects of animism–Buddhism (Visalo 2003: 176). The Thai middle class has been drawn to three new religious movements: Buddhadasa Bhikku's, Santi-Asoke's, and Dhammakaya's.

Buddhadasa Bhikku revisited the Paticca Samuppada model or the Dependent Origination, which is interpreted as a cycle of physical birth and rebirth in another form of life after one's death. It provides a new translation of the original Mahanidana Sutta in Pali (palisuttas.com 2015), i.e., the birth and rebirth of me, I, and mine on every moment of thought (nkgen.com 2012). Once we can get rid of "atta" (the perception of self) to become "anatta" (the perception of non-self), then we will reach Nibbana (in Pali) or 5. Nirvana (in Sanskrit) or sunyata (voidness of kilesa or suffering), which is a state of supreme bliss .

Santi-Asoke is a Buddhist sect that declared itself as separated from the mainstream Thai Hinnayana Buddhism in 1975 (Asoke 2008). The Buddhist monks of Santi-Asoke call themselves "samana" and adhere to austere practices emphasizing sustainability in producing organic food, self-reliance, sacrifice, 5 Buddhist Precepts, especially abstinence from gambling and intoxication, and living a simple life (comparable to the Amish communities in the USA) (Asoke 2008).

Last but not least, there is the *Dhammakaya* sect, which has drawn many disciples and built up its one billion US dollar worth mega projects, such as the colossal Dhammakay Cetiya, the grandiose cloister, and the unorthodox Assembly Hall which looks like a spaceship or even a UFO (amusingplanet.com 2016). The pilgrimage should be done in a jungle, not in an urbanized area. Walking on flower petals that believers spread on the street for monks to step on (like walking on a red carpet) is not what Buddha taught (Guinnessworldrecords.com 2015). It has been known for its outlandish marketing strategies to draw donations and financial support from devotees (manager 2015). The sermons of *Dhammakaya* are controversial because it contradicts the dogma of anatta in Theravada Buddhism or Hinnayana Buddhism. Veneral P.A. Payutto or Phra Bhramagunabhorn analyzed the Dhammakaya teachings and concluded that Dhammakaya's teaching is not in line with neither Theravada nor Mahayana Buddhism (Payutto 2008a). Moreover, Dhammakaya practices its own rites and rituals which deviate from those of mainstream Buddhism.

Dhammakaya has its own Websites and telecommunication means to connect with devotees. Via its digital communication networks, Dhammakaya has gained more popularity among its followers, but at the same time, it has aroused anti-Dhammakaya movements online as well. On January 5, 2016, Patcharawalai Sanyanusin, a writer for the *Life* section of the *Bangkok Post*, reports about Phra Dhammachayo, abbot of Wat Phra Dhammakaya:

> …While I don't think it's reasonable that one should turn their backs on the clergy because they're unhappy with some of its members, I can't help but wonder how certain members who have intentionally committed much more serious offences are still embraced by their followers and the clergy.
>
> The most apparent example is Phra Dhammachayo, abbot of the controversial Wat Dhammakaya who successfully, yet disgracefully, escaped being disrobed for embezzlement, which he was charged with 17 years ago. He's the country's most infamous and influential monk who has long been at the forefront of many scandals. If any of these incidents had involved ordinary monks they would have been expelled from the monastic community almost immediately. The temple has also long been a target of criticism as it earns massive wealth from its greed-promoting donation concept.
>
> However, the temple's worst offence, which has long been condemned by many Buddhist scholars and revered monks, is its attempt to fuse its own theories into Buddha's teachings and spread them far and wide.
>
> Phra Brahmagunabhorn (widely known as Than Payutto), who is locally and internationally recognized for his profound expertise in *Tripitaka* (Buddhist canon), warns that distorted lessons don't only cause Buddhist followers to misunderstand the traditional principles, but also damage the religion at its roots.
>
> After Wat Dhammakaya teachings started spreading in early 1999, the reserved scholar wrote a well-documented book to explain all subjects that the temple discretely twisted.
>
> But the biggest question that has troubled many concerned Buddhists from the beginning is why the monkhood' governing body never came up with a resolute measure to stop the temple's hideous malpractices. When Phra Dhammachayo was accused of embezzling 900 million baht in assets from his temple in 1999, he was never held accountable. At that time, the late Supreme Patriarch Somdet Phra Nyanasamvara instructed the Supreme Sangha Council to defrock him for stealing and distorting Buddhism, but it did nothing as many of its members sided with him.
>
> Finally, the monk returned the assets to the temple and walked away, and the council immediately closed the case and refused to reopen it. It's hard to believe that many high ranked monks are too blind to see the damage being spread by Phra Dhammachayo and the temple instead of using their authority to punish them and fix a growing problem, they turn a blind eye to it.
>
> When news broke that we're going to have a new Supreme Patriarch soon, I was positive that this problem might be given serious consideration. But, I was disappointed at the revelation that more than half of the eight highest ranked priests in the council, who are likely to be candidates for the top job, are thought to have good connection with Wat Dhammakaya.
>
> I don't know whether their relationships have anything to do with its wealth, but I can only say that my little hope is now seeing a dimmer light at the end of the tunnel (Sanyanusin 2016).

The latest news that put Wat Phra Dhammakaya on the front page of newspapers is a scandal about the embezzlement of the credit union funds at Khlong Chan in 2014. It was reported as a court case and an investigation from the Department of Special Investigation (DSI) on Wat Phra Dhammakaya about a multi-million Baht donation from the former President of the Khlong Chan credit union Cooperative to both the abbot and the temple. The question was whether the donation was to be considered as a credit union customers' investment (Bangkok Post 2015a, Nationmultimedia 2015). Finally, in 2015, the Dhammakaya Temple negotiated the return of 684 million Baht back to the credit union in exchange for dropping the lawsuit (Matichon TV on YouTube 2015). The DSI investigation is still ongoing, even though the credit union dropped the charge.

As Thai people are becoming more and more interconnected due to the advance of digital communication, texts and image sharing on certain controversial issues go viral at times. Recent statistics (tech.thaivisa.com 2015) indicate that, of the total of 64.9 million Thais, active Internet users in Thailand account for 23.9 million (37% of the population), active social media accounts 32 million, mobile connections 97 million (mobile Internet users are 17.7 million or 27% of the population), and active mobile social accounts 28.0 million. The Website also reports that Thai active Internet users spend on average 5.5 h a day on their PC or tablet, and 4 h a day on average on their cell phones. Thais spend on average 3 h and 46 min on social media and watch about 2 h and 46 min spent of television.

Thai Internet users represent the middle class, the "target audience" of the Dhammakaya marketing strategies. Realizing that Internet users do not represent the entire population, this author is still interested in investigating how the Thai middle class receives the Dhammakaya and how Dhammakaya has withstood all the critiques and public scrutiny over the years.

2.2 Objectives of This Study

The objectives are twofold: first, to describe general information related to Dhammakaya online; and second, to critically investigate the discussion online about the Dhammakaya case by using qualitative content analyses.

This study is based on the academic analysis of the worldview, value, symbolic representation, and social organization of Wat Phra Dhammaya via information available online.

2.3 Research Questions

(1) How are Wat Phra Dhammakaya and the abbot perceived among online users?
(2) How is the online Dhammakaya case a good case in point of hybridization of the Thai religious culture as a result of the politico-economic influences in the contemporary globalization period?

2.4 Methodology

Internet research was undertaken from March 25, 2015, till January 6, 2016. A Google search using specific keywords such as "Dhammakaya" and "ธรรมกาย" was executed. All news online about Dhammakaya—the temple, doctrine, abbot, scandals, and opinions—posted in the cyber world since 2013 was categorized and analyzed.

A Twitter hashtag search for keywords "Dhammy Dhammakaya" and "Naja Dhammakaya" was undertaken in the same time frame. One anti-Dhammakaya Website, one Twitter account, and eight Facebook pages were studied. Discussion about Wat Phra Dhammakaya on pantip.com and blogs were investigated. Also, one pro-Dhammakaya Website was investigated. Interviews of famous monks and Dhammakaya analyses of P.A. Payutto, V. Vajiramedhi, and Phra Paisal Visalo, who are three intellectual Hinnayana Buddhist monks in Thailand, were studied.

2.5 Findings and Discussion

Quantitative content analysis of Dhammakaya search online

From Table 2.1, we can categorize online information related to Dhammakaya Temple into three types: facts, opinions, and news.

Examples of *facts* are the biography of the abbot (Dhammachayo Bhikku), information about Vijja Dhammakaya (the core of Dhammakaya which is a meditation technique), and facts about Dhammakaya founded by Luang Por Sod (born 1884), a famous monk from Wat Pak Nam Phasi Charoen (Kom Chad Luek 2015).

Opinions and comments are what follows: (1) criticism about the abbot's lifestyle, his claimed supernatural abilities, such as claiming to know where Steve Jobs' ghost resides after his death, and whether the abbot is qualified to be a Buddhist monk; (2) negative comments and opinions from scholars, famous monks, and readers on the temple. invasive marketing strategies to gain donations from the public and the accountability and transparency of the temple; (3) 17

Table 2.1 Results of "ธรรมกาย" (Dhammakaya) keyword search on Google (research undertaken on March 27, 2015 and December 26–29, 2015)

Facts	Biography of abbot	Text	003
		Clip	001
	Vijja Dhammakaya (core of Dhammakaya)	Clip	024
	Facts about Dhammakaya	Text	015
Comments/opinions	Criticism on the abbot	Text	002
		Clips	012
	Negative opinions on the temple from readers/scholars/monks	Text	018
	Anti-Dhammakaya	Websites/Webpages/blogs	017
	Positive comment	Text	001
News	News about activities of the Wat (Temple) Phra Dhammakaya	Text	062
		Clip	003
	News about questionable accountability and transparency of the administration related to the temple and the abbot	Text	056
		Clip	004
	News released from Wat Phra Dhammakaya	Text	009
		Clip	001
Total			216

anti-Dhammakaya online pages (Facebook, Webpages, Websites, and blogs); and (4) one positive comment on the temple. Only one out of forty pieces of comments/opinions is positive.

News about Wat Phra Dhammakaya can be identified as (1) general news about the activities of the temple; (2) news on dubious accountability related to the temple and the abbot during the past scandals and the credit union cooperative at Khlong Chan, which is a current one; (3) and news released from the temple.

From the results, 97 out of 216 pieces of information (44.9found. On Facebook %) are negative.

From a rough estimation of the *comments/opinions* online, around half of the information referred to the former Prime Minister Taksin Shinawatra and his followers. That means that the temple was said to be linked with a populist political party.

Dhammy is the colloquial term by which many Thai Internet users refer to the abbot as it is derived from his title, "Dhammachayo". From Table 2.2, a hashtag search indicates how Wat Phra Dhammakaya has been on Twitter. Three types of information were found. First, *facts*: the biography of the abbot. Second, *comments/opinions* of which 122 pieces are negative toward the abbot and/or the temple. To elaborate, 35 pieces of information were about the personal appearance and lifestyle of the abbot; 50 pieces of information were critiques on the claimed supernatural ability of the abbot (such as seeing Steve Jobs' life after death, offering food to the Lord Buddha in Nirvana, and describing how holy-beings in heaven are

Table 2.2 Results of googling #ธัมมี ธรรมกาย (Dhammy Dhammakaya), researched on March 27, 2015

Facts	Biography of abbot	Text	008
Comments/opinions	Criticism on the abbot	Text	035
	Negative opinions on the temple from readers/scholars/monks	Text	037
	Negative opinions/VDO clips criticizing the supernatural ability of the abbot	Text/clips	050
	Sermons of the abbot (about rebirth)	Clips	024
News	News about activities of the Wat (Temple) Phra Dhammakaya	Text	007
	News about questionable accountability and transparency of the administration related to the temple and the abbot	Text	031
	News released from Wat Phra Dhammakaya	Text	005
	News about monk status of the abbot	Text	004
Total			193

dressed); and 37 pieces of information were about the temple, such as the criticism of the ability of Nun or Mae Chee Chan Khonnokyoong, whom the abbot reveres, to cast away the nuclear bomb which was directed toward Thailand to Japan during the Second World War. Third, *news*: 31 out of 40 news pieces were about the questionable accountability and transparency relate to the administration of the temple and the abbot regarding the credit union Cooperative at Khlong Chan case.

"Na ja" is the suffix with which the abbot likes to end his sentences when he speaks to his followers, in a similar way an adult would talk to a child. Such a keyword yields much information about Dhammakaya.

From Table 2.3, this hashtag search indicates three types of information: facts, comments/opinions, and news. One fact is about the biography of the abbot, but the other four pieces are about the history of the temple.

About 70 percent of the comments (110 out of 153 comments) are negative. Mostly, negative opinions are on the core teaching of Dhammakaya (or Vijja Dhammakaya), which is, in fact, a meditation technique.

Among the news related to Wat Phra Dhammakaya, 28 out of 34 pieces of news were related to the questionable accountability and transparency of the administration of the temple and the abbot.

Anti-Dhammakaya comments and opinions online are posted on several different social networks such as Facebook, Twitter, Google Plus, and blogs. Also Websites that launched negative criticism toward the temple were found. On Facebook, the page "*Join us to fight against Wat Phra Dhammakaya and stop materialism for the status quo of Buddhism*" (2015) consists of images and explanations why this page is against this temple. Next is the Anti-Dhammakaya Federation page (2015). This page asks the temple to stop selling merits (*bun*, in Thai) to pupils and teachers and

Table 2.3 Results of googling # นะจ๊ะธรรมกาย (Na ja Dhammakaya); research undertaken on March 27, 2015

Facts	Biography of abbot	Text	001
	History of the temple	Text	004
Comments/opinions	Criticism on the abbot	Text	004
	Negative opinions on the temple from readers/scholars/monks	Text	011
	Negative opinions/VDO clips criticizing the supernatural ability of the abbot	Text/clips	007
	Disclosure of Vijja Dhammakaya (negative)	Text/clips	092
News	News about activities of the Wat (Temple) Phra Dhammakaya	Text	001
	News about questionable accountability and transparency of the administration related to the temple and the abbot	Text	028
	News released from Wat Phra Dhammakaya	Text	001
	News about monk status of the abbot	Text	004
Total			153

stipulates that the core of Dhammakaya did not exist in Buddha's teaching. Other pages are *"Revelation of Dhammakaya"* (2015) which criticizes the behaviors of the abbot as not in conformity with those of a Buddhist monk and that Dhammakaya is a cult; and "Unmasking Dhammakaya" (2015) which disagrees that the abbot of Wat Phra Dhammakaya acts as an astrologer. Another page is called "X-Phra Dhammakaya" (2012). It follows the lawsuit cases of the temple and comments on the information released by the temple that the preaching of the temple does not conform to the Tripitaka (the Buddhist canon in Pali). Moreover, it criticizes the advertisement of chances to get jewelry in heaven if one donates to the temple. It also displays photographs of important politicians, members of the cabinet, and high-ranked government officials who attend ceremonies of the temple. This affirms that the temple has good relations with the polity of Thailand.

A Twitter account found is the *"Anti-Wat Phra Dhammakaya"* (2015). This account posts links to update the followers with the lawsuit case.

Also Websites that are against the performances of Wat Phra Dhammakaya can be found. See, for instance, Coolcial.com (2015). The main aim of this Website is to advocate to Thai youth to disagree with the teaching of the temple. Another one is DMC (2015). This Website allows the anti- and pro-Dhammakaya members to discuss intellectually. One Website that accused the temple of getting involved with politics is alittlebuddha.com (2015).

Also worth mentioning are online media blogs, such as the campaign against Dhammakaya on *Thai Post* online (2015). Blogs that are against Dhammakaya can be listed as (1): "I hate Dhammakaya" (forum.uamulet.com 2015); and (2) "Why are we against Dhammakaya" from the Pantip Webboard (2015). Offline campaigns against Dhammakaya were reported in Ranong (esanguide.com 2015). That incident happened when Ranong residents protested against the blocking of roads at Wat Phra Dhammakaya for its march on March 2, 2015 (*Bangkok Post* online 2015b).

From contents online, Wat Phra Dhammakaya shares similarities with other Thai Buddhist temples that sell supranatural power (of amulets, of the abbot, of the meditation technique, etc.). It induces more believers to make merits for one's own wealth gain. These phenomena manifest the Thai hybridized religious culture explained in Chap. 1. However, the differences displayed by Wat Phra Dhammakaya are (1) the promotion of Nibbana as atta (self); (2) the promotion that the abbot is the prophet; (3) the massive wealth gain (multi-million Baht) of the temple such as the encroachment of land of the poor to expand the temple and the invasive marketing strategies to sell merits; (4) the good relations with the Phue Thai political party and the members of the Sangha Council, especially the Acting Supreme Patriarch; and (5) as related to (4), the ability to escape huge financial scandals and the abbot from being disrobed.

Let us start from what makes Dhammakaya notorious online.

(1) The promotion of Nibbana as atta: Teaching of the abbot and messages from Wat Phra Dhammakaya online showed that Nibbana (Pali spelling) or Nirvana (Sanskrit spelling) is a land where Lord Buddha lives after he passed away and the abbot could go there in his meditative body to offer food to Lord Buddha. There is a ritual to offer food to the Lord Buddha at this temple.

Discussion online on the *Pantip website* searched on March 28, 2015. Pantip is a famous discussion forum in the Thai cyberspace. Findings reveal that online users are questioning whether Nibbana or Nirvana or the state of supreme bliss is *atta* (in Pali, atman in Sanskrit—meaning personality, ego, soul, or self), according to Wat Phra Dhammakaya's claim, or not.

2.6 Analysis

Payutto (2008b: 95–103) stated clearly that teaching about atta does not conform to what Buddha taught. Actually, there is nothing we can hold on to as we all age, go through phases of change, get sick, and eventually die, either of old age, sickness, or accidents. The world keeps changing and nothing is here to stay. Impermanence or transience (*anicca*) is evident. Buddhists know that the Buddha taught that there is "no self" (*anatta*), and that the doctrine of *anatta* has become a dogma and a component of Buddhist identity (May 1984: 93).

The clinging and attachment to "self" can be explained by the doctrine of causal genesis or 3.Dependent Origination (*Paticca Samuppada*). It can be traced through 12 conditioned factors that we fabricate our "self." Buddhadasa Bhikkhu (1906–1993) explains the 12 factors as the birth of I and mine (or self) causes ignorance to the truth that the world is transient. That ignorance breeds 10 kinds of unwholesome qualities of the mind or *kilesa*: greed (*lobha*), hate (*dosa*), delusion (*moha*), conceit (*mana*), speculative views (*ditthi*), restlessness (*uddhacca*), shamelessness (*ahirika*), Doubt (*vicikiccha*) and sloth (*thina*) and lack of moral dread or un-conscientiousness (*anottappa*). That mental formations feed consciousness. Consciousness or *vinnana* is a part of the five aggregates or *Panca-khandha* which give us the individual illusory of self (or *atta*). The other four aggregates are as follows: matter (*rupa*), feeling (*vedana*), perception (*sanna*), and mental formation (*sankhara*). The five aggregates constitute six-based self-impression: eyes, ears, nose, tongue, skin, and mind. Those organs make sense contact: vision, hear, smell, taste, touch, and perception. Those six senses cause feelings. Then, those feelings cause craving for sensory cognizable objects, which cause attachment or clinging to sensed objects. The clinging to sensed objects feeds the coming into being of self. That causes the birth of "I" and "mine." This will go on in our trains of thought ceaselessly.

Buddhadasa Bhikkhu (1906–1993), who is revered as the most influential Buddhist philosopher in Thailand, and Phra Brahmagunabhorn or Bhikkhu P. A. Payutto, who is widely regarded as the living expert on Thevarada Buddhism in Thailand, studied Suttapitaka as the source of doctrinal authority on different occasions. They both refuted the interpretation of the PS model as the cycle of past, present, and future life or re-becoming (Jackson 2003: 90–91). Bhikkhu (2002: 20) stated clearly that "Paticcasamuppada (in his own spelling) is a matter of the highest ultimate truth; it is not a matter of morality. There is no self which travels from life to life and no need to say that one cycle of PS must cover three lifetimes, as understood in the language of relative truth."

Bhikkhu (2002: 19) affirmed that "this life means the cycle of 3.Dependent Origination; the next life means the next cycle of 3.Dependent Origination, and so on." The cessation of self can happen any moment of practice in our lifetime. From his arduous Pali studies and interpretation of many ancient books, he came up with a new interpretation as seen in the following chart:

2.7   The 12 Causal Links in the PS Model

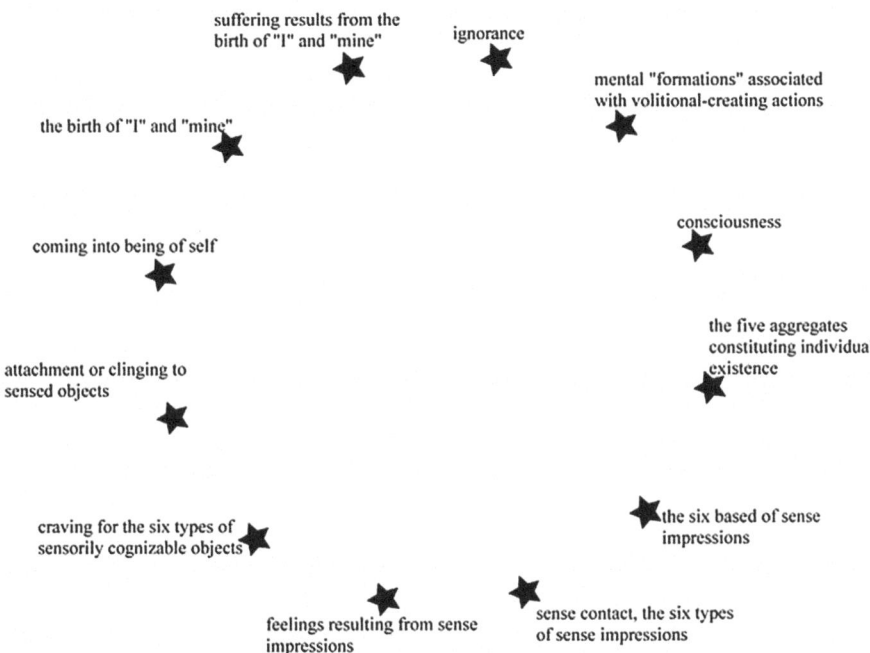

suffering results from the
birth of "I" and "mine"   ignorance

mental "formations" associated
with volitional-creating actions

the birth of "I" and "mine"

coming into being of self   consciousness

the five aggregates
constituting individual
existence

attachment or clinging to
sensed objects

craving for the six types of   the six based of sense
sensorily cognizable objects   impressions

feelings resulting from sense   sense contact, the six types
impressions   of sense impressions

Diagram constructed by Patchanee Malikhao from Jackson (2003:111–119).

"It [PS Model] is a detailed demonstration of how suffering arises and how suffering ceases. It also demonstrated that the arising and ceasing of suffering is a matter of natural interdependence. It is not necessary for angels or holy things, or anything else to help suffering to arise or cease…. The other aspect of [PS Model] is that it demonstrates that there are no sentient beings, persons, selves, we or they here or floating around looking for a next life. Everything is just nature: arising, existing, passing away" (Buddhadasa 2002: 23–24).

This interpretation of Buddhadasa boldly contradicts the traditional interpretation, which has widely mistaken the birth of I, myself, and mine (*atta*) with the birth of a human life and the suffering from the birth of I, myself, and mine (*atta*) with the old age, sickness, and death. Buddhadasa famously stated that "Nirvana here and now," meaning that one can attain Nirvana in this lifetime as one can be trained to be a selfless person. The ways to the cessation of "the subjective sense of self (or *atta*) together with the self-centered attitudes associated with it," as Jackson (2003: 116) states, were explained by Buddha as the *four Noble Truths* (Abe 1989: 205–296, Jackson 2003: 132,142, Thomas 1951: 96).

The PS model shows that the becoming of self and the passing away of self are impermanent as long as the mind moves.

From this explanation, atta (or self) does not exist in Buddhism. Therefore, there is no atta to argue about the state of Nibbana or Nirvana (Payutto 2008a: 105–106).

(2) The promotion that the abbot is the prophet: This is what online discussants, especially on pantip.com (2015), criticized about. A page on Google Plus is called "*Stop Dhammakaya*" (หยุด!ธรรมกาย 2015). It equates the Dhammakaya as a cult. It criticizes the behavior and clothing of the abbot. Moreover, it denounces the teaching that does not conform to the Buddhist canon.

2.8 Analysis

The claim that Abbot Dhammachayo is a Phra Ton-Thad-Ton- Dham (the Origin of Elements and Dhamma) (Laohavanich 2016a: 5) made Theravadian scholars criticize that Dhammakaya uses the Mahayana Buddhism in the form of Theravada Buddhism for its own financial gain (Scott 2009). Beliefs in Mahayana Buddhism can be explained that all sentient beings can achieve Buddhahood and all have to work compassionately to release other's sufferings (Buddhist tourism.com 2007). The Ton-Thad-Ton- Dham notion is not quite likely in line with the Mahayana's beliefs, but in fact, it is in line with the Vajarayana Canon of Tibet which emphasizes the battle between the Army of Light and the Sons of Darkness or Mara and supernatural power (Laohavanich 2016a: 5, 2016b). Dhammakaya already offers different ways of rites and rituals other than those of mainstream Thai Theravada Buddhism, such as the pilgrimage in the urban area, offering food to Buddha in Nirvana land, and procession of Visakha the female disciple of Buddha. The Vijja Dhammakaya or the core of Dhammakaya offers a different path from how Lord Buddha used to attain enlightenment. It is a meditation technique that Luang Por (Rev.) Sod from Wat Pak Nam Phasi Charoen rediscovered; it is a visualization meditation technique (Scott 2009: 147–148). This technique is one of the 40 plus tranquil meditation techniques known before the emergence of Theravada Buddhism. Lord Buddha was the one who discovered the meditation technique called "insight meditation" or "anapanasti—breathing meditation" to kill the kilesa or defilements completely by reflecting upon the trilakkana (dhukka—suffering, anicca—impermanence, and anatta—non-self) of all beings (Budsas.org 1988). However, the abbot of Wat Luang Por Sodh Dhammakayaram, which is a separate temple from Wat Phra Dhammakaya but originated from Wat Pak Nam Phasi Charoen, claims that Vijja Dhammakaya meditation technique is an insight meditation as well through the visualization and wisdom technique (Phra Rajyanvisith[1] 2011: 36–137).

---

[1] 1 The abbot of Wat Luang Por Sodh Dhammakayaram in Ratchaburi Province.

(3) Massive wealth gain, land encroachment, etc: Even though Wat Phra Dhammakaya is not much different from the majority of Wats in Thailand that offer magic monks, amulets, holy figurines and statues, mantra, tattoos, etc., the temple was accused of its invasive marketing technique via online and mass communications of the temple. Online users discussed whether the marketing strategies of Wat Phra Dhammakaya conform to what Buddha taught, i.e., the more money one donates to the temple, the more merit points one will gain in life; paying-by-installment merit making to the temple is also accepted; and selling heaven to donators. Furthermore, the temple has had many lawsuit cases because villagers were evicted or the temple encroached on public land for its own expansion.

2.9 Analysis

Mediatization and commercialization of Thai Buddhism, as a result of modernization in the contemporary globalization period explained by Malikhao (2015: 124–143), have continued to feed the sense of self. Malikhao (2012, 2015) reports that Thai religious culture has been affected by the media in a long run (also known as the mediatization process of Thai religious culture), which can be seen in the proliferation of cults, practices, and beliefs. The perception of self feeds instant gratification according to the Paticca Samuppada wheel. Then, craving and lust or wanting to be or wanting not to be arises.

The Thai hybridized religious culture, analyzed by using Servaes' (1999: 12) four interrelated analytical components: a worldview (*Weltanschauung*), a value system, a system of symbolic representations, and a social-cultural organizational system, reveals that Wat Phra Dhammakaya Temple manifests the outlook of modernity but the content is the same old cocktail animism–Brahmanism–Theravada Buddhism. Cleanliness and orderliness of the temple, as a symbolic representation of modernity, can draw many middle-class believers. Marketing strategies to draw in public to make merit by advocating on amount of merit points one gets in life depends on the amount of money one donates to Dhammakaya, the portrayal of heaven, organized activities such as mass pilgrimage, etc. are the "brand" that the temple has been creating for the past 30 years. As Ekachai (2016) reports about the branding image of Dhammakaya:

> It focuses on orderliness, cleanliness, and grandeur. This strikes a chord with the middle class and the new rich who believe in supernatural powers but want a temple with a modern look and style to suit their worldly status. The fund-raising groups also give followers a sense of community in a big city, not to mention the business connections that come with it. In short, Dhammakaya answers the needs which the irrelevant clergy fails to do.

Another point of criticism is about the pictures on the catalogue, "the spell-bound charm of heaven," printed by Wat Phra Dhammakaya about jewelry-adorned accessories the donators will receive when they die and go to heaven (Dhammachayo 2005). The abbot also talked about these accessories on television (Phu Jud Karn 2015). In this regard, stated Phra Paisal Visalo, an intellectual Thai Theravada monk, this kind of merit making does not happen to polish our greed and cravings at all. To the contrary, it creates more greed and craving. This kind of teaching does not conform to what Buddha taught (Visalo interview in Prachatai.com 2015).

Phra Paisal Visalo (Phu Jud Karn 2015) in his interview with the Phu Jud Karn newspaper indicates that merit-making messages from Wat Phra Dhammakaya do not conform to what Buddha taught.

> The right kind of dana in Buddhism (Suppurissadana) comprises (1) giving away clean things; (2) giving away delicate things; (3) giving away at the right time; (4) giving away an appropriate gift; (5) giving away with conscience; (6) giving away often; (7) when giving away, the mind will be crystal clear; (8) after giving away, the mind will be blissful. By the way, giving away material things or dana is only one kind of merit making. Many kinds of merit making do not require money at all, such as adhering to the Five Precepts (abstain from killing, lying, stealing, committing adultery, drinking intoxicants), praying or meditating, listening to sermons/Dhamma, being modest, having a right view. All of the aforementioned merits are much greater than giving away money.

But giving away money to temples, instead of organization for the good course, seems to be habits of Thais in general. This belief that one can accumulate merits to become a richer person in the next life or can be traced back in the Mahapadana sutta or Dhammapada Nisaya, the text used in Northern India (Scott 2009: 29–31). Wat Phra Dhammakaya is not different from other popular temples, but it amplifies this value with the branding of "the more you give, the more you reap in your next life."

Wat Phra Dhammakaya's giving away amulets or Buddha image inscribed as "Dood Sap" (inducing assets) to big donors is a symbolic representation of the animistic element of Thai religious culture. Wealth gain of the temple through donation and merit making are part of the sociocultural system of the Thai religious culture. "Gaining wealth if you donate more" is already established as the world-view of popular Thai Buddhists. Wat Phra Dhammakaya just dares say it obviously and promotes these beliefs through their communication channels: the Internet, TV, radio, mailing lists, merit representatives, etc.

However, what Wat Phra Dhammakaya advertised through its own media about the supranatural ability of nuns from this temple who could fly to brush off the atomic bomb heading to Thailand to Japan during the Second World War (2Bangkok.com 2002), and that of the abbot himself who claimed that he could meet Steve Jobs after his death in heaven (2Bangkok.com 2012) and could go to offer food to Lord Buddha in the Nirvana land (Scott 2009: was ridiculed and severely criticized online and in the main stream mass media. These messages come directly from the temple, not from the followers or the mainstream media. That put Wat Phra Dhammakaya in the category of promoting a wrong worldview which was not approved by Lord Buddha.

(4) Good relations with the Phue Thai political party and the members of the Sangha Council, especially the Acting Supreme Patriarch:

Thai Buddhism institutions have undergone a reform to be under the state control since the reign of King Rama V (Scott 2009: 59). The 1962 Act is a symbolic representation of the hierarchical value that puts the Sangha Council (Buddhist monk body) to govern the monks with the Supreme Patriarch at the top of the chain of command (Scott 2009). Thai Buddhist monks have ranking and titles deferring to the central authority from Bangkok.

Wat Phra Dhammakaya was known to have a good relationship with the former Thai Rak Thai (TRT) party or now Phue Thai (PT) party of the deposed Prime Minister Thaksin Shinnawatra. According to the Nationmultimedia.com (2006), the leaders of the political party and the Temple are both shrewd investors, bold to take new risks, fond of modern technology, modernity, and capitalism. Mr. Thaksin was praised by the Temple as a good example of the person who made good merits in the past life and, therefore, he became a billionaire in this life and the Temple began to support a candidate from the TRT party since 2000 (Nationmultimedia.com 2006).

Wat Phra Dhammakaya belongs to the Mahanikaya of Theravada Buddhism. The relationship between Wat Phra Dhammakaya and Wat Pak Nam Phasi Charoen was dated back since the ordination of Abbot Dhammachayo whose preceptor is the current Somdet Phra Ratchamangalajarn, the abbot of Wat Pak Nam Phasi Charoen. Monks from Mahanikaya have been invited to attend or preside over ceremonies organized by Wat Phra Dhammakaya.

(5) As related to (4), the ability to escape huge financial scandals and the abbot from being disrobed:

The National Reform Council's (NRC) committee on religious affairs stated that that Abbot Dhammachayo violated the Buddhist monks' code of conduct (Patimokkha) (Bangkokpost.com 2015c). It cited a letter, dated on April 26, 1999, written by the late Supreme Patriarch, His Holiness Somdet Phra Nyanasamvara, who listed two infringements by Abbot Dhammachayo: (1) distortion of Buddhist doctrines which causes conflicts between Buddhist monks; and (2) reluctance to transfer 1500 rais of land which was donated to him by his followers to Wat Phra Dhammakaya (Bangkokpost.com 2015c). The letter was sent to the Supreme Sangha Council (SSC), but the SSC did not defrock abbot Dhammachayo. Again for the scandal of the Khlong Chan credit union, the SSC affirmed on February 10, 2016, that it would not consider pursuing the Dhammakaya case (Thairath 2016).

The Dhammakaya controversy reflects the crisis within the weak Supreme Sangha Council that tolerates any deviation from teaching and practicing Theravada Buddhism. Dhammakaya belongs to the Mahanikaya Order of Theravada Buddhism but it does not promote the core of Theravada Buddhism, such as the Buddhist dogma of anatta. The Sangha did not rule that Dhammakaya is a deviant of Theravada Buddhism. [Another order of Theravada Buddhism is the stricter Dhammayutti-nikaya Order, established during the reign of King Mongkut (Rama IV); the Dhammayutti monks dress themselves differently and are stricter in their

Dhamma practice but they adhere to the same Buddha's teaching as the Mahanikaya].

Dhammaya is a good example of networking between the Supreme Sangha Council, high-ranked monks from provinces, a political party, politicians, high-ranked civil servants, and multi-millionaires/investors. Ekachai (2016) in her article, entitled "Supreme Patriarch Row won't help Clergy" in *Bangkok Post online*, states clearly how Dhammakaya has withstood all the frictions and controversies:

> The temple's close connections with the elders explain why the Sangha Council did not follow through with the late Supreme Patriarch's ruling against Phra Dhammachayo, on claims of divisive teaching and theft, which could have led to his defrockment.
>
> Its close connections with political and business elites also explain why so many lawsuits against Dhammachaya on public fraud never stick. It is no secret that the key candidate, Somdet Phra Maha Ratchamangalacharn, the abbot of Wat Paknam, is close to Dhammakaya. But so are several other members of the Sangha Council.

The wealthy Dhammakaya has no problem pampering the elders who also view Dhammakaya's expansion overseas as a global expansion of Thai Theravada Buddhism without the Sangha having to lift a finger. Dhammakaya scholarships to monks over the years have also expanded the movement's support base nationwide. If the next Supreme Patriarch is a Dhammakaya supporter, it is feared the controversial sect will take over the whole Sangha. Distorted Buddhist teaching will be institutionalized and the allocation of the much sought-after clerical ranks will be also decided by Dhammakaya, giving it total control over the clergy.

Phra Paisal Visalo (Phu Jud Karn 2015) affirms what Ekachai reports in his interview that Dhammakaya (affiliated to the Mahanikaya Order) has been inviting senior monks from the Mahanikaya Order and members of the Supreme Sangha Council to its ceremonies for the past 30 years. The relationship is good, and they have common interests. No wonder that those monks have a positive attitude toward the temple. Currently, Somdet Phra Maha Ratchamangkalacharn, from the Mahanikaya Order, is acting Supreme Patriarch and that would give more "soft power" to the temple.

Dhamakaya was counted as one of the morality boost camps by Prachatai.com (2015), which include Santi-Asoke, Col. Chamlong Srimuang (former Governor of Bangkok), Phra Payom Kalayano, and 264 more organizations. These groups were united in protesting against the Thai Beverage Company, the brewer of Chang Beer, to be registered as a public company. Prachatai also speculated that the conservative or morality camps will support the revision of the 2007 draft constitution (especially measures 9, 21, and 22 to strengthen penalties for those who violate Buddhism, Buddhist leaders, or Buddhist institutions). Prachatai argues that these will violate the right of expression of the people in criticizing Buddhism.

2.10 Conclusion

To answer the two research questions, first, Wat Phra Dhammakaya and the abbot were perceived as negative among online users. The temple was perceived as influential and linked with a major political party and the members of the Supreme Sangha Council. As a result, the temple and the abbot survived scandals and accusations. The relationship between the temple and politics, including the Sangha politics, indicates that Wat Phra Dhammakaya possesses "soft power" which makes it withstand criticism and avoid court cases.

Dhammakaya is a good case to study the hybridization of the Thai religious as a result of politico-economic influences. Marketing strategies, such as enticing the public to donate more and more for heavenly life after death, or persuading the public to make merits on the mass pilgrimage in the urbanized areas, and so on, are what the temple has been implementing for the past 30 years. This is obviously a merging of consumerism and Thai Buddhism. Hence, the second research question is answered.

Well noted that when the Phue Thai party was in power, many lawsuits against Dhammakaya were dropped. At the time of writing this article, under the new military government, the DSI is still investigating the case of Dhammakaya, even though the SSC did not want to do anything.

References

Abe, M. (1989). Zen and Western Thought (W. R. LaFleur, Trans.). Honolulu, HI: University of Hawaii Press.

Bhikku, B. (2002). *Paticcasamuppada Thammasapa, Bangkok* (1st ed.). Bangkok: Thammasapa Publishing.

Dhammachayo, L. P. (2005). *The spell-bound charm of heaven (มนต์เสน่ห์แห่งสวรรค์).* Pathumtani: Dhammakaya Foundation.

Ekachai, S. (2001). *Keeping the faith: Thai Buddhism at the crossroads post books.* Bangkok: The Post Publishing Plc.

Formoso, B. (2000). *Thaïlande: Bouddhisme Renoncant Capitalisme Triomphant: La Documentation Francaise* (p. 00001). Paris: JSTOR.

Jackson, P. (2003). *Buddhadasa, Theravada Buddhism and modernist reform in Thailand.* Chiang Mai: Silkworm Books.

Kitiarsa, P. (2012). *Mediums, monks, & amulets Thai Popular Buddhism today.* Chiang Mai: Silkworm Books.

Laohavanich, M. M. (2016a). Thai Theravada Buddhists and digital technology. A paper presented on "Religions in Digital Asia" on the Asian Round Table discussion, St. John's University, Bangkok, February 16th.

Laohavanich, M. M. (2016b). Oral Presentation on Thai Theravada Buddhists and digital technology, St. John's University, Bangkok, February 16th.

Malikhao, P. (2012). Sex in the village. Culture, religion and HIV/AIDS in Thailand. Penang-Chiang Mai: Southbound-Silkworm books.

Malikhao, P. (2015). Thai Buddhism, the mass media and culture change in Thailand. *Journal of the Asian Research Center for Religion Communication, 12*(2), 124–143.

Payutto, P. A. (2001). Buddhism in contemporary Thailand". In *Thai Buddhism in the Buddhist world. A survey of the Buddhist situation against a historical background*. (pp. 137–154). Bangkok: Buddhadhamma Foundation.

Payutto, P. A. (2008a). กรณีธรรมกาย บทเรียนเพื่อศึกษาพุทธศาสนาและสร้างสรรค์สังคมไทย (Dhammakaya case: A lesson for Buddhism studies and creative Thai Society). Nakorn Pathom: Bodhidham Association. http://www.watnyanaves.net/uploads/File/books/pdf/the_ Dhammakaya_case_lesson_learned_for_buddhist_education_and_society_development_ expanded and revised.pdf. Accessed January 4th, 2016.

Payutto, P. A. (2008b). Karanee Dhammakaya (Dhammakaya Case) (24th ed.). Bangkok: Bodhidhamma Association.

Rajyanvisith, P. (2011). *The heart of Dhammakaya Meditation* (4th ed.). Ratchaburi: Wat Luang Por Sodh Dhammakayaram.

Prachachart, S. (2003, 21st–27th October). 10,000-million amulet business: The world most expensive mass. Matichon Weekly, 23, 13.

Sanyanusin, P. (2016). Crime and no punishment. The Bangkok Post (Life Section), January 6th.

Scott, R. M. (2009). *Nirvana for sale? Buddhism, wealth, and the Dhammakaya temple in contemporary Thailand*. New York: State University of New York.

Servaes, J. (1999). *Communication for development: One world, multiple cultures*. New Jersey: Hampton Press Inc.

Thomas, E. J. (1951). *The history of Buddhist thought*. London: Routledge & Kegan Paul.

Visalo, P. (2003). *Thai Buddhism in the future: Trend and ways out of crisis (in Thai language)*. Bangkok: Sodsri-Saritwong Foundation.

Visalo, P. (2015). "พระไพศาล วิสาโล" "ธรรมกาย" เป็นที่เชิดหน้าชูตาของ "มหานิกาย" (Interview of Phra Pisan Visalo: Dhammakaya is the crown jewel of Mahanikaya). ASTV Phu Jud Karn daily, March 7th, 2015.

Websites

Alittlebuddha.com.   (2015).   http://www.alittlebuddha.com/News2014/February2014.html. Accessed on Janauary 6th, 2016.

Amusingplanet.com. (2016). http://www.amusingplanet.com/2014/01/the-magnificent-buddhist-temple-of-wat.html. Accessed on January 6th, 2016.

Anti-Dhammakaya   Federation.   (2015).   https://www.facebook.com/AntiDMCFederation. Accessed on January 5th, 2016.

Anti-PraDhammakaya. (2015). https://th-th.facebook.com/AntiPraDhammakaya. Accessed on January 5th, 2016.

Anti-PhraDhammakaya—Twitter.   (2015).   https://twitter.com/   atDhammakaya1,   or ต่อต้านวัดพระธรรมกาย (@ATDhammakaya1) | Twitter Accessed on January 5th, 2016).

Asoke. (2008). http://www.asoke.info/Book/santi_history/santiasoke_history.html. Accessed on December 27th, 2015.

Bangkok Post. (2015a). http://www.bangkokpost.com/archive/dsi-finds-b600m-was-transferred-to-wat-phra-Dhammakaya/669032. Accessed on January 1st, 2016.

Bangkok Post. (2015b). http://www.bangkokpost.com/archive/ranong-pushes-out-Dhammakaya-march/486002. Accessed on January 5th, 2016.

BangkokPost. (2015c) http://www.bangkokpost.com/learning/learning-from-news/478960/phra-dhammachayo-charges-of-corruption-renewed. Accessed on March 4th, 2016.

Bangkok Post. (2016). http://www.bangkokpost.com/opinion/opinion/817320/. Accessed on January 6th , 2016.

Baan Lan Siang Dham or nkgen.com. (2012) http://www.nkgen.com/739.htm. Accessed on December 26th, 2015.

Budsas.org. (1988). http://www.budsas.org/ebud/ebmed012.htm. Accessed on March 3rd, 2016.

Buddhadasa Bhikku (2002). *Paticcasamuppada Thammasapa, Bangkok* (1st edn.). Bangkok, Thakland: Thammasapa Publishing.

Coolcial.com. (2015). http://www.coolcial.com/th/fanpage/155320694517077/feed. Accessed on January 6th, 2016.

DMC. (2015). http://www.dmc.tv/forum/index.php?showtopic=12197, Access on January 6th, 2016.

Education Management Information System Centre. (2016). http://www.moe.go.th/main2/article/e-hist01.htm. Accessed on January 5th, 2016.

Ekachai, S. (2016). Military, Sangha share many similarities. http://www.bangkokpost.com/opinion/opinion/890864/military-sangha-share-many-similarities. Accessed March 11th, 2016.

Esanguide.com. (2015). http://www.esanguide.com/travel/detail.php?id=6036. Accessed March 27th, 2015.

Forum.uamulet.com. (2015). http://forum.uamulet.com/view_topic.aspx?bid=2&qid=2658. Accessed on January 6th, 2016.

Guinness World Records. (2015). http://www.guinnessworldrecords.com/world-records/longest-journey-walking-on-flower-petals. Accessed December 28th, 2015.

Kom Chad Luek 2015. (May 31st, 2015). http://www.komchadluek.net/detail/20150531/207200.html. Accessed January 1st, 2016.

Manager Online. (July 2nd, 2013). http://www.manager.co.th/Local/ViewNews.aspx?NewsID=9560000079619. Accessed on December 26th, 2015.

Manager Online. (February 27th, 2015). http://www.manager.co.th/mwebboard/listComment.aspx?Mbrowse=8&QNumber=383090. Accessed on December 27th, 2015.

Matichon TV via YouTube. (March 16th, 2015). https://www.youtube.com/watch?v=AUuEvKe1u1k. Accessed on January 1st, 2016.

Nationmultimedia.com. (September 27th, 2006). http://www.nationmultimedia.com/2006/09/27/opinion/opinion_30014752.php. Accessed on March 4th, 2016.

Nationmultimedia.com. (March 22nd, 2015). http://www.nationmultimedia.com/national/Monks-confirm-getting-big-cheques-from-former-coop-30256512.html. Accessed on January 1st, 2016.

Palisuttas.com. (2015). http://palisuttas.com/2015/01/25/mahanidana-sutta-dn-15-2/. Accessed on December 26th, 2015.

Pantip.com. (2015). http://pantip.com/topic/30730423 Accessed March 27, 2015.

Pantip Web board. (2015). http://dog508dod.blogspot.hk/2012/01/blog-post_04.html. Accessed on March 27th, 2015.

Prachatai.com. (2015). http://www.prachatai.com/journal/2012/04/39991. Accessed January 4th, 2015.

Stop Dhammakaya. (2015). https://plus.google.com/communities/104133698005331003169. Accessed on January 5th, 2016.

Tech.thaivisa.com. (2015). http://tech.thaivisa.com/complete-insight-internet-social-media-usage-thailand/3147/#prettyPhoto. Accessed on January 22nd, 2016.

2 Bangkok.com. (2002). http://2bangkok.com/2bangkok-news-7361.html. Accessed on March 4th, 2016.

2Bangkok.com (2012). http://2bangkok.com/Dhammakaya-knows-jobs-afterlife.html. Accessed on March 4th, 2016.

Unmasking of Dhammakaya. (2015). https://th-th.facebook.com/pages/แฉธรรมกาย/254150517955549. Accessed on January 5th, 2016.

Unmasking Dhammakaya (แฉธรรมกาย) (2015). https://www.facebook.com/แฉธรรมกาย-254150517955549/. Accessed on January 5th, 2016.

Thai Post. (2015). http://www.ryt9.com/s/tpd/2104716. Accessed on March 27th, 2015.

Thairath. (2016). http://www.thairath.co.th/content/575663. Accessed on March 4th, 2016.

X-Phra Dhammakaya (2012–2016) https://www.facebook.com/pages/X-วัดพระธรรมกาย/429431607072760. Accessed on January 5th, 2016.

Chapter 3
Violence Against Thai Females

Abstract This chapter discusses violence against females from a Thai historical perspective. The Thai hierarchical worldview, especially toward females, will be explained. Rape, domestic violence, verbal abuse, spiritual violence, sexual harassment, bullying, trafficking, child pornography, etc., which are the symbolic representations of violence against female rights, will be analyzed. The highlight of this chapter is the discussion about portrayals of violence against females in the mass media.

3.1 Introduction

Violence against women is a common phenomenon in many countries. It is estimated that 35% of women worldwide have experienced some sort of violence. This could be dating violence, domestic and intimate partner violence, emotional abuse, human trafficking, same-sex relationship violence, sexual assault and abuse, stalking, violence against immigrant and refugee women, violence against women at work, and violence against women with disabilities (womenshealth.gov 2015). To elaborate, the Website womenshealth.gov (2015) gives detailed information of all sorts of violence that can be summarized as follows:

Dating violence against women happens when a female experiences the following from her date: physical abuse such as hitting, shoving, kicking, biting, or throwing things; emotional abuse such as yelling, name-calling, bullying, embarrassing, keeping one away from one's friends, saying one deserves the abuse, or giving gifts to make up for the abuse; sexual abuse such as doing something sexual without her consent in sobriety.

Domestic violence against women is when a male in a relationship commits physical, emotional, or sexual abuse to the woman who is in a relationship with him.

Emotional abuse against women is when a female is experiencing monitoring of what she is doing all the time, is accused of being unfaithful all the time, is discouraged to see friends or family, is barred to go to work or school, is frightened when the partner shows anger, is controlled how she is spending money, is humiliated in front of others, etc.

Women trafficking is when a woman is tricked or forced into working in a terrible condition.

Same-sex relationship violence.

Violence against females occurs in many cultures. Many scholars state that violence against females is a form of power that masculinity exercises over femininity. And that can be traced back in history. In Thai culture, it is known that the culture is hierarchal and patriarchal throughout the history of Thailand.

3.2 Female Status in Thai Culture from a Historical Perspective

From historical evidence, from the ancient Indian time, women were under the guardian of their parents when they were young, of their husbands when they were married, and of their sons when they had children. When the husband passed away, the widow needed to stay single for the rest of her life or, before the English colonizer forbade, they jumped into the pyre used for cremating the husband's dead body in public (Bhikkhu Mettanando 2004: 214–224). During the Lord Buddha's time, Lord Buddha allowed women to be ordained as bhikkhunis or female Buddhist monks. Evidence that bhikkhunis preached to bhikkhus (male Buddhist monks) are the Songs of the Elders, a literature translated from Pali into English from the Thera and Theri Gather as a part of Tripitaka (the three collections of books making up the Buddhist canon of scriptures—dictionary.com 2016) (Bhikkhu Mettanando 2004: 235). However, Bhikkhu Mettanando (2004: 212, 248–252, 274–275, 283) researched that discrimination against the bhikkhunis process occurred after the death of Lord Buddha. Dhammacaro (n.d.) reports oppression of bhikkhunis by imposing the eight Garudhammas (rules for bhikkhunis to adhere to) which are as follows:

1. "However old a bhikkhuni may be, she must pay respect even to a newly ordained monk and should learn and practice this dhamma throughout her life.
2. A bhikkhuni must not stay in a nunnery to observe the Buddhist Lent where there is no bhikkhu Mettanando nearby and should learn and practice this dhamma throughout her life.
3. A bhikkhuni must invite a bhikkhu Mettanando every fortnight to fix the date of the Sabbath and the day to listen to the exhortation (Ovada) of the monks and should learn and practice this dhamma throughout her life.

4. A bhikkhuni must perform the ceremony of confession and taking advice both in the bhikkhu Mettanando Sangha and the bhikkhuni Sangha and should learn and practice this dhamma throughout her life.
5. A bhikkhuni must observe the manattna discipline, first from a bhikkhu Mettanando and then from a bhikkhuni and should learn and practice this dhamma throughout her life.
6. A bhikkhuni, after training in six pacittiya rules of the bhikkhuni patimokkha, should seek upasampada from both bhikkhu Mettanando and bhikkhuni sanghas and should learn and practice this dhamma throughout her life.
7. A bhikhhuni must not admonish a bhikkhu Mettanando and should learn and practice this dhamma throughout her life
8. Since having become a nun, she should be receptive to learning and should learn and practice this dhamma throughout her life."

The assumption is that female discrimination in Buddhism, resulting in the extinction of bhikkhunis, was the reason that Buddhism disappeared from India in the 11th century and reappeared in Sri Lanka, while Jainism has been in existence in India because this religion supports female Jainist monks (Bhikkhu Mettanando 2004: 280–283). Bhikkhu Mettanando (2004: 282–283) argues that women played an important role in propagating Buddhism as women payed more attention in the life course of people; the destruction of bhikkhuni status is a consequence of Brahmanism which believes that women are dirty and will bring shame on the family.

I tend to disagree with Mettanando Bhikkhu's explanation that the inequality between genders among Buddhist monks was the reason that Buddhism disappeared in India. The social context of India has not supported gender inequality since the ancient time, whether or not the social context was under the influence of Islam or Hinduism. Women were reported to be the victims of domestic violence in many states in India in the 2000s and many girls are married at a young age (Prasad 2008: 32–33). Moreover, in India many serious crimes of raping women and hurting or killing them came into international attention. One explanation that Buddhism disappeared in India was that Lord Buddha was integrated into an incarnation or an avatar of Vishnu in Hinduism and therefore, there is no need to have a separate religion in India (The Times of India 2001).

Discrimination against women in the Thai Buddhist culture can be seen from the historical perspective as follows: During the archaeic globalization (before 1500s), Buddhism came into pre-Siam (the former Thailand) via Sri Lanka after 1291 (Ishii 1986: 60). According to Beasley (1999: 72), misogynist literature suggests that femininity obstructs the Symbolic Order or Law of the Father's rules. Sripariyattimoli (1998: 54) argues that the literature may have been influenced by Hindu literature which emphasized male supremacy. Although, as explained earlier, there were some elements of female subordination among Buddhist monks, the status of female laities was noticeably affected when Brahmanism mixed with Buddhism. Males were and still are the ones who perform rites and rituals, a practice influenced by Brahmanism (Sripariyattimoli 1998: 49). Females, it was argued, menstruate and as a result are considered unclean and are therefore forbidden from participating in special religious ceremonies and entering sacred sites organized by men (Srivanichpoom 2004; Kawanami 1996: 74–75).

During the Proto-Globalization period (1600–1768), the *sakdina* system[1] was introduced reinforcing the lower status of females (Baker and Phongpaichit 2005: 16–17). Females in the lower class were subjected to the sakdina lords. It was a tradition that the royals had many wives to ensure the production of enough sons to assist with administrative tasks and enough daughters to build marriage networks within the elite. According to Wyatt (1984), Ayutthaya kings had many wives.

During the Globalization period (1768–1946), the hierarchical culture was strengthened by King Rama V's modernization reform. Bureaucracy established in this period replaced the sakdina system from the Ayutthaya period, but the continued presence of a patriarchal society and polygamy ensured the retention of this status quo in practice (Baker and Phongpaichit 2005). Male supremacy went so far that females and children in the middle and lower class were treated as goods before 1868. Evidence was the Act of 1868 which stopped males from selling their wives and parents from selling their children (Achawanijkul and Tharawan 2005: 272).

According to Formoso (2000: 65), King Rama VI decreed a polygamy forbidding Act in 1921 but it was not imposed rigorously. An Act to endorse monogamy was approved some time later in 1935 when a group of middle-class females asked for an equal chance to education and monogamy (Achawanijkul and Tharawan 2005: 335). However, the monogamy value seemed to lose ground to the hybridized form of sexuality emerging in this Globalization period increasing acceptance of having sex with sex workers. As polygamy for males was still accepted, while females had to stay chaste before marriage, males turned to a third party: sex workers (Ghosh 2002). Ghosh (2002: 34–35) argues that prostitution was officially mentioned for the first time in Thai history during the reign of King Rama I when prostitution and brothels were taxed, a levy called "tax for the road". This implied that prostitution was treated and accepted as a profession. Jeffrey (2002: 11) explains the sex industry was initially booming after the complete abolishment of slavery by King Rama V in 1905 when the former slave wives entered prostitution to survive. The number of sex workers was on the rise after the Great Depression in 1936 (Jeffrey 2002: 15). Klausner (1997: 67–69) and Sivaraksa (2001: 36–37) blame the Thai patriarchal system as a stimulus for prostitution. Poor young Thai males can get free education and accommodation from a temple up to university level and can afterward disrobe to work as laity, whereas poor young females have no institution support and end up being sex workers.

During the Contemporary Globalization (1946–Present) period, patriarchal values continue. Eoseewong (1992: 41–42), a famous Thai scholar, explains that in the past Thai females played an important role in the production of the households such as working in paddy fields, harvesting, sending food to the males who

---

[1]The sakdina system was the hierarchical structure of service nobility codified in lists of official posts, each with its specific title, honorific and rank measured in areas of land they were allowed to possess (Baker and Phongpaichit 2005: 15; Ongsakul 2005: 115; Servaes and Malikhao 1989: 33; Servaes 1999: 211; Srisootarapan 1976; Suwannarit 2003: 9–12).

ploughed the fields, and growing field crops such as peanuts. The females were the ones who inherited assets from the parents. However, the status of the females has been considered lower than that of the males in two aspects: First, in Buddhism, the females are not allowed to listen to a sermon from a Buddhist monk in the front row, to hand food directly to the monk's hand, and to be the masters of Buddhist ceremonies. Second, governance, males are the head of the household. Females still do the housework.

In this period, the Vietnam War took place. The USA had seven air force bases in Eastern Thailand (Baker and Phongpaichit 2005: 149). From these bases, US war planes left to bomb North Vietnam and Laos. Bangkok became the hub of the GIs R&R (rest and recreation) tours. Bars, nightclubs, brothels, and massage parlours mushroomed in Bangkok and around the air force bases. The number of sex workers in Bangkok increased to up to 300,000 (Baker and Phongpaichit 2005: 149). In the 1960s, migration of women and girls from the North into the sex industry followed. The northern region has a distinctive culture that reinforces prostitution, as Jeffrey explains:

> In the North, where animism remained strong in spite of the predominantly Buddhist character of the country, women acted as guardians of the spirit. Sexual relations before marriage were viewed as an offence to the family spirits, which could be ameliorated through an offering of gifts from the couple to the woman's familial spirits in a phit phi (wronging of the spirits) ceremony. Some analysts now view this particular ceremony as one of the links between women's sexuality and monetary value that has made prostitution more acceptable in the North (Jeffrey 2002: 30–31).

By 1980, when the tourism industry started to flourish in Thailand, the number of sex workers was between 500,000 and 700,000 (Jeffrey 2002: 78). The neoliberal impact on the economy affected lower class women significantly. The International Monetary Fund's Structural Adjustment Program (SAP) and the General Agreement of Tariffs and Trade (GATT) prescribed the forms of increasing liberalism, privatization, and deregulation of economies in many poor countries. As a result, small entrepreneurs were submerged by monopoly capitalists (Ghosh 2002: 49). Women's labour became exploited by monopoly capitalists in a way that could be called modern-day slavery (Ghosh 2002: 50–53). Eoseewong (1992: 44–45) reports when capitalism spread to the rural areas, Thais experiences the breakups of families and communities as parents leave home to get a harvest job or cutting sugar canes for the industry for a long while. Grandparents take care of the children. Teachers live in the urban areas and ride on motorcycle to teach in the village. Thus, there is no participation in the village. Headmen (phu yai ban) and head of tambon (kamnan) or the local leaders became government officials. They are no longer part of the locals. Worse than that, the commodification of the women's bodies and women trafficking occurred. Eoseewong (1992: 47) states that selling daughters is a sign of the downfall of the female status.

Beyrer (1998: 28–29) suggests the high value placed on female virginity and male sexual freedom also reinforced the concept of exploitation of women from a lower class. Sittirak (1996) argues that the process of capital accumulation, which Thais adopted from the West in the 1950s, is similar to that adopted in the colonized

Third World. Western capitalist patriarchy cannot be achieved without maintaining the oppression and exploitation of women by men. However, Western capitalist patriarchy has blended with Thai patriarchy to create a hybridized form of Thai prostitution whereby most females view themselves as family breadwinners (Sittirak 1996; Boonchalaksi and Guest 1998).

Ekachai (bangkokpost.com 2016) states on the International Women's day (March 9) that both the clergy and the military are extremely authoritarian and feudal, patriarchal, and sexist. Ekachai explains the way women have to avoid physical touching to monks, because the clergy would consider this a sexual temptation, as a cultural affirmation of women's inferiority. Ekachai indicates that the fact that women are not allowed to be ordained as bhikkhunis is a sign of being oppressed. (Some women went to have a cross ordination from the Mahayana Canon in Taiwan but those female monks are not accepted by the Thai Theravada Sangha on the claim that Theravadian bhikkhunis who must be preceptors in the ordination in Theravada Buddhism went into distinction. Therefore, they are not counted as bhikkhunis.) Ekachai (bangkokpost.com 2006) states, "For the Thai Sangha, female ordination is a no-no. Female monks are treated as illegal, and monks who support it are subjected to banishment. The white-robed nuns are accepted because they accept their inferior status as temple help and never as monks' equals." To elaborate, a bhikkhuni or a female monk has to adhere to 311 precepts and wears yellow robe as a male monk but a white-robed nun has to adhere to only 8 precepts. That is why Ekachai concludes that Thai women are losers in the patriarchal system.

In fact, the right and status of females has been improved by law substantially. The subcommittee for female role and activity development (1981: 195, 200–201) reports that after the revolution in 1932, the first constitution and the first election act allowed both males and females to have equal political right: as candidates and as voters. In 1952, females could become judges. In 1970, the civil law prohibits marriage registration for more than once. Wives can choose to work and manage certain assets gained from marriage and can get alimony. Importantly, if the husband acknowledges other women as wives, the registered wife can sue for a divorce. The Ministry of Defence allowed females to be ranked higher than Lieutenant Colonel. The Ministry of Internal Affairs also allowed females to be ranked higher than a deputy sheriff, and they can be governors. In 2003, females can choose to use their last names or their husbands' last names (Institute of Public Policy Studies 2014). However, in practice gender inequality in the Thai society still exists and persists.

Malikhao (2012: 185–189) studied 14 focus group discussions on current sexual norms, values, and understanding of HIV/AIDS. She found that most students interviewed think that being a male has more sexual privilege than being a female, even though males and females have equal opportunities to study and find work. Although sex education has been taught in schools in Thailand for about three years by the time of the study in 2011, gender inequality still persists. Malikhao (2012: 187) reports:

Most male and female students under study think that rape cases occur because young women wear too sexy clothes. When asked why those who wear normal clothes were raped, the students argued that this was because the rapists got sexually aroused from seeing those girls who wear sexy clothes … and afterwards when there is a chance of not being caught, they just rape any woman available … most students view the violence on Thai TV as normal. And

Sitcoms or soap series which portray a male main character kissing a female main character who then slaps his face, or a male character who rapes the female character out of jealousy and then falls in love with the victim later, can be watched on a daily basis.

3.3 Analysis: Gender and Power as Represented in the Media

"Thailand has a social sphere that considers rape cases normal phenomena, as TV and plays portray and support the rape culture. The general public may think that these (portrayals) cause nothing in their sub-consciousness; they have no reactions, nor do they question whether it is normal that the main characters perform in a rape scene before they become a pair. We can see that this rape culture is deep-rooted. There are, as well, advertisements that may indicate a degree of violence against women but we think that they are okay; they are just advertisements", stated Ms. Tasanawan Banjong, host of Divas Inter on Voice TV and administrative staff of the Friedrich Ebert foundation (Thailand) (Wiprawit 2016a).

This above-mentioned statement about the "rape culture" may seem exaggerating, had we not delved into sexual violence cases in Thailand. According to Khaosod English newspaper online (March 8, 2016), shocking statistics unveiled by the Ministry of Social Development and Human Security, together with, the Women and Men Progressive Movement Foundation (WMP) that an average of 87 cases of sexual violence are reported each day in Thailand, and that means one case per 15 min. A record of 31,866 cases in 2013 is terrifying; the victims were children and females of which the youngest was a one-year-and-nine-months-old girl and the oldest was an 85-years-old woman (Khaosod 2016). Seminars, campaigns, and discussions have been organized in Thailand to call for a change in the portrayals of children and women in the media. For instance, a campaign to stop displaying the rape scenes in any drama or series on TV was undertaken in 2015 and that lead to a petition to stop the idea that rape and seduction for rape was normal (Wiprawit 2016b). This leads to a call for reviews of scripts and collaborations between TV program producers and the universal services obligation, which controls the broadcasting and telecommunication of Thailand, on zero-rape scenes on TV dramas and series. Melodrama is a portrayal of the collective relationship in a community. The assumption that violence seen on television may cause violence in real life is based on the hypodermic theory that the mass media have direct effect on the audience, which has been proven not true. However, we can see that what people like on TV is about violence on gender. This reinforces the patriarchal worldview of Thais.

According to the Website rabbit daily (2016), ten popular drama series are all about the male main characters raping the female main characters:

(1) Saneha sanya kan (affection and revenge). It is about the male main character who lost his just married wife in an accident. He thought that the female main character was the murderer. He, therefore, pretended to be good with her and then raped her until she became pregnant.

(2) Rang Ngao (shadow power). The female and male main characters had a quarrel, and the male raped the female.

(3) Sawan Biang (skewed heaven). The male main character did not like the elder sister of the female character for so many reasons, and he then raped the female character. She became pregnant. Later, they fell in love.

(4) Dao Phra Suk (Venus). The female main character was forced to be a prostitute. She fell in love with a customer (a male main character who already had a fiancée). She wanted to quit the relationship, and he then raped her.

(5) Fai rak asoon (demon's fervent love). When the wife of the main character disappeared, he was looking for her. He found that she was going to marry another man. He then dragged her home, chained her, and raped her. It turned out that she was a twin of his wife. Later, the woman who got raped had a happy ending with the man who raped her.

(6) Game rai game rak (mean game, love game). The male character raped the female character who lost her memories, but he did not know the facts.

(7) Roi Leh Saneh Luang (a hundred tricks, a seduction). As the dad of the female main character raped the mom of the male main character, the male main character then raped the female out of revenge.

(8) Jamloei Rak (victim of love). The sister of an innocent female main character used her name to commit a crime. The male main character then kidnapped the female to an island. Slapped and kissed and raped her. Later, they fell in love.

(9) Tawan Tor Saeng (shining sun). The male and female characters, who are in a relationship, had a quarrel. Then, the male raped the female.

(10) Koo Karm (karma couple). The main male character was drunk and raped his wife, the female main character.

Another kind of melodrama rated as popular among the audience is the kind that two or three female characters catfight both physically and verbally over a male character. What does this mean? Why do Thais consider physical and verbal violence and rape a kind of entertainment?

Many researchers on Thai society confirm that the socioeconomic change in Thai society has not affected the tradition and sexuality of Thais (Archawanitchakul and Tarawan 2005: 80–81). The roles of females and males as portrayed by the media are still very much in conformity with the mainstream discourse. That is males are the leaders in family and society. They have better administrating jobs and better education while females are the followers or practitioners. The females' jobs can be in the service industry, employees in companies, healthcare services, etc. Moreover, the males in advertisements are lively, vigorous, and active while the females are shy, dare not express their sexuality in conformity with the mainstream discourse that good females must keep their virginity, and know little about sex. This discourse differentiates "good women", those who keep their virginity, from "bad women", those who sell their bodies. The study of Manokasemsook et al. (2008: 171–258) on sexuality on TV advertisement found that television advertisements in Thailand are made to reinforce the sexual norm that females lack negotiation power, and, as a consequence, suffer. The third gender is not accepted in advertisements. Moreover, Thai advertisements stigmatize those who do not conform to the sexual norms.

Chanthong et al. (2008: 67–112) studied sexuality of Thai teenagers online via the camfrog program. It is a social media network equipped with a video camera on which female members show their nude image live and male members watch and comment. They found that many females disguised themselves by using a male username. The male usernames use vulgar and sexist language. This is a conflict of power where the female showed that they had power by choosing to show or not to show their bodies to seduce the males. At the same time, the male commentators use their abusive language to show that they also had power above the image. The females use a male's name to conform to the sexual norm that good Thai women do not express or talk about sex.

Vulgar and sexist language can be found in general comments from male users online on social media networks. Gossiping and mudslinging are what we sometimes see from Web boards that allow the online users to raise issues for discussion. Although, traditional Thai folk plays, such as choi or pleng yao, also contain under the belt language and dirty jokes or pun words, the verbal violence online and bullying is now growing as we can see from many Websites and anti-Websites that anyone create by using a false name to criticize a person online. These phenomena are symbolic representations of the conflict avoidance value of Thais.

3.4 Conclusion

It is undeniable that the Thai worldview is patriarchal, and that can be traced back for centuries. There are those who argue that Thai culture gives honor to females as many important compound words have the word, "mae" or mother in them, such as Mae Nam or river. Inheritance was given to the daughters. Traditionally, when a male and a female got married, the male moved to build a house in the female's land. But during the contemporary period, that does not happen anymore due to social change. Urbanization has changed the family unit to a nuclear family. The price of land has increased, and people in the urban areas do not have enough space to build another house for their daughters. More and more couples live in a small condominium. More important than that the worldview of Thais is still the same as in many cultures: masculinity is dominant. Symbolic representations can be seen from the polygamous behavior of Thai males, rape as entertainment in melodrama, and sexist comments found online. Sexy portraits of girls are on the front pages of popular newspapers. Selling cars or selling beer needs young women in sexy clothes to present the goods. Also social organization as seen in Thai Buddhism, and females are still not allowed to be monks. Females are not allowed to enter some holy spaces in temples.

My concern is that sex education being taught at school cannot change the attitude of students that 'raping is okay because the females gesture in a seductive way by dressing themselves so sexy'. It is good to know that some civil groups launched a petition to stop violence on the mass media and go against promoting rape unconsciously. I do think that sex education curricula need to be reassessed and readjusted together with the control and censorship on violence against females in the mass media.

References

Archawanijjakul, K., & Tharawan, K. (2005). Karnmauang Rauang Paed lae Rangkai Phooying: AIDS Taeng Kwamroenrang lae Ying Rak Ying (Sex and female body politics: AIDS, abortion, violence and women who love women). In A. Pongsapich (Ed.), *Paed Sathana lae Paedwithee nai Sangkom Thai (Gender and sexuality in Thai society)* (pp. 269–339). Bangkok: Chulalongkorn University Press.

Baker, C., & Phongpaichit, P. (2005). *A history of Thailand.* New York: Cambridge University Press.

Beasley, C. (1999). *What is feminism anyway?.* St Leonards: Allen & Unwin.

Beyrer, C. (1998). *War in the blood: Sex, politics and AIDS in Southeast Asia.* Bangkok: White Lotus.

Bhikkhu Mettanando, M. (2004). *What happened in B.E. 1 (Haet Kerd Por Sor Neung).* Bangkok: Sahassawat Foundation.

Boonchalaksi, W., & Guest, P. (1998). Prostitution in Thailand. In L. L. Lin (Ed.), *The sex sector: The economic and social bases of prostitution in Southeast Asia* (pp. 130–169). Geneva: International Labour Office.

Chanthong, W., Boonmongkol, P., Samakkikarom, R., & Manokasemsuk, C. (2008). Camfrog and sexuality of Thai female teenagers. In P. Boonmongkol et. al. (Eds.), *Sexuality in popular media: Risky context and supporting context to sexual health (Pedwithi nai Sue Niyom: Boribot Saeng lae Serm Tor Sukkaphawa Tang Ped)* (pp. 69–112). Bangkok: Foundation for Female Sexual Health and Institute of Social and Population Research, Mahidol University.

Dhammacaro. (n.d.). Female Monks in Buddhism. http://www.buddhapadipa.org/buddhism/female-monks-in-buddhism/. Accessed March 10th, 2016.

Ekachai, S. (2016). Military, Sangha share many similarities. http://www.bangkokpost.com/opinion/opinion/890864/military-sangha-share-many-similarities. Accessed March 11th, 2016.

Eoseewong, N. (1992). Status of women: Past, present, and future. In *Karn Prachum Smatcha Haeng Chart (Report of National Council Conference).* Bangkok.

Formoso, B. (2000). *Thaïlande: Buouddhisme Renoncant Capitalisme Triomphant.* Paris: La Documentation Francaise.

Ghosh, L. (2002). *Prostitution in Thailand: Myth and reality.* New Delhi: Munshiram Manoharlal Publishers Pvt. Ltd.

Ishii, Y. (1986). *Sangha, state, and society: Thai buddhism in history* (P. Hawkes, Trans.). Honolulu: University of Hawaii Press.

Jeffrey, L. A. (2002). *Sex and borders: Gender, national identity and prostitution policy in Thailand.* Chaing Mai: Silkworm Books.

Kawanami, H. (1996). Women in buddhism revisited. In T. Cosslett, A. Easton, & P. Summerfield (Eds.), *Women, power and resistance*. Buckingham-Philadelphia: Open University Press.

Klausner, W. (1997). *Thai culture in transition*. Bangkok: The Siam Society.

Manokasemsook, C., Boongmongkol, P., & Samakkikarom, R. (2008). Sexuality on TV advertisement. In P. Boonmongkol et. al. (Eds.), *Sexuality in popular media: Risky context and supporting context to sexual health (Pedwithi nai Sue Niyom: Boribot Saeng lae Serm Tor Sukkaphawa Tang Ped)* (pp. 171–258). Bangkok: Foundation for Female Sexual Health and Institute of Social and Population Research, Mahidol University.

Malikhao, P. (2012). *Sex in the village. Culture, religion and HIV/AIDS in Thailand*. Penang: Southbound and Silkworm books.

Ongsakul, S. (2005). *History of Lan Na (C. Tanratanakul, Trans)* (1st ed.). Chiang Mai: Silkworm Books.

Prasad, K. (2008). Women's vulnerability to HIV/AIDS in Asia and Africa. In K. Prasad, & U.V. Somayajulu (Eds.), *Empowering women worldwide series: 1. HIV and AIDS. Vulnerability of women in Asia and Africa* (pp. 1–57). Delhi: The Women Press.

Servaes, J. (1999). *Communication for development: One world, multiple cultures* (1st ed.). Cresskill, New Jersey: Hampton Press Inc.

Servaes, J., & Malikhao, P. (1989). How 'culture' affects films and videos in Thailand. *Media Development, 36*(4), 32–36.

Sittirak, S. (1996). *Daughters of development: The stories of women and the changing environment in Thailand*. Bangkok: Women and Environment Research Network in Thailand (WENIT).

Sivaraksa, S. (2001). *Manodharmsamneuk samrab Sangkom Ruamsamai (Conscience for contemporary society)*. Bangkok: Religion for Development Committee.

Sripariyattimoli, P. (Kusalcitto, S.) (1998). *Women's status in buddhism*. Bangkok: Maha Chulalongkorn Rajawittayalai Press.

Srisootarapan, S. (1976). *Chom Nar Sakdina Thai (The face of Thai Sakdina)*. Bangkok: Agsorn Sampan.

Srivanichpoom, M. (2004, 16th–22nd July). Women are not allowed to enter. *Siam Rath Weekly*, 51.

Suwannarit, V. (2003). *Withi Thai (Thai ways)*. Bangkok: Odian Store.

Wiprawit, N. (2016a). Rape from the monitor to reality? https://www.change.org/p/nbtcupdate-เลิกเผยแพร่คติการล่อลวงข่มขืนว่าเป็นสิ่งปกติ-supinya/u/7693905. Accessed March 8th, 2016.

Wiprawit, N. (2016b). Attended the seminar, "The portrayals of children and women on Thai TV" https://www.change.org/p/nbtcupdate-เลิกเผยแพร่คติการล่อลวงข่มขืนว่าเป็นสิ่งปกติ-supinya/u/8168695. Accessed March 8th, 2016.

Wyatt, D. K. (1984). *Thailand: A short history*. New Haven and London: Yale University Press.

Websites

Dictionary.com. (2016). http://www.dictionary.com/browse/tripitaka?s=t. Accessed March 9th, 2016.

Institute of Public Policy Studies. (2014). http://www.fpps.or.th/news.php?detail=n1059126642. news. Accessed April 15th, 2016.

Khaosod English online. (2016). http://www.khaosodenglish.com/detail.php?newsid=1395313389&section=12&typecate=06. Accessed March 8th, 2016.

Rabbit daily. (2016). https://daily.rabbit.co.th/. Accessed April 15th, 2016.

The Times of India, May 7th, 2001. http://timesofindia.indiatimes.com/articleshow/40860434.cms. Accessed March 9th, 2016.

Womenshealth.gove. (2015). http://www.womenshealth.gov/violence-against-women/types-of-violence/dating-violence.html. Accessed March 11th, 2016.

Chapter 4
A Village in the Jungle: Culture and Communication in Thailand

Jan Servaes

Abstract This article provides analytical components of Thai culture: worldviews, value systems, symbolic representation, and social organization. The Thai Buddhist worldview and beliefs will be discussed from a historical perspective. Power hidden in the hierarchical structure will be assessed in the realm of interpersonal and mass communication. Thai cultural products will be analyzed in relation to the Thai value system and symbolic representations. How Thais organize their social life will be illuminated by looking at the role and impact of 'face' and 'leadership' in Thai society.

There appears to be an almost insulting contradiction between the image of the delicate Land of Smiles, of exquisite manners and 'unique hospitality', and the world of live pussy shows. Yet, to see these images as contradictory is perhaps to misunderstand Thailand. Patpong kitsch and Thai traditions coexist—they are images from different worlds, forms manipulated according to opportunity. The same girl who dances to rock 'n' roll on a bar top, wearing nothing but cowboy boots, seemingly a vision of corrupted innocence, will donate part of her earnings to a Buddhist monk the next morning, to earn religious merit. The essence of her culture, her moral universe outside the bar, is symbolized not by the cowboy boots, but by the amulets she wears around her neck, with images of Thai kings, of revered monks, or of the Lord Buddha. The apparent ease with which Thai appear able to adopt different forms, to swim in and out of seemingly contradictory worlds, is not proof of a lack of cultural identity, nor is the kitsch of Patpong proof of Thai corruption—on the contrary, it reflects the corrupted taste of Westerners, for whom it is specifically designed. Under the evanescent surface, Thais remain in control of themselves.

Buruma (1989: 30)

4.1 Introduction

Though Thailand has never been "colonized" by a foreign power, external influences have always played a significant role in the country's history. Whereas the Chinese and Indian cultures heavily affected the local society historically, since the beginning

The original version of the chapter was revised: Missing author name has been updated. The erratum to the chapter is available at 10.1007/978-981-10-4125-9_10

This is an updated and revised version of Chap. 8, published in Servaes (1999), Communication for Develop7ment. One World, Multiple Cultures, Cresskill NJ: Hampton Press, pp. 209–225.

of the so-called Ratanakosin period in 1782, Western values were introduced. Especially during the reigns of King Chulalongkorn or Rama V (1868–1910), King Vajiravudh or Rama VI (1910–1925), and the regime of Colonel Pibul (1938–1945), the Thai were strongly "advised," not to say "forced," to assimilate through "fashion," language, and etiquette, the forms and symbols of a so-called Western civilization: "Material progress, technical advancement, and high standard of living characteristic of the West make the Thai think that Western culture must be better than ours and that it is our duty to follow suit and adopt it as ours." This statement is made by the late Kukrit Pramoj, who, as a former Prime Minister, founder and president of the Social Action Party, actor in the film "The Ugly American" (with Marlon Brando), newspaper columnist, author of several historic novels, and expert in the traditional *Khon* pantomime art, can be considered one of the outstanding representatives of Thai culture. He concluded that "it must be admitted that Thai culture is in a state of utter confusion, and probably it has reached the highest degree of confusion ever known in our history" (Pramoj, in Van Beek 1983: 93–94).

How to research or comprehend this complexity? Obviously, we first of all have to come to grips with our past. Said's (1985) captivating overview of the way in which Asian societies and philosophies throughout the ages were perceived by the West starts from the thesis:

> That the essential aspects of modern Orientalist theory and praxis (from which present-day Orientalism derives) can be understood, not as a sudden access of objective knowledge about the Orient, but as a set of structures inherited from the past, secularized, redisposed, and re-formed by such disciplines as philology, which in turn were naturalized, modernized, and laicized substitutes for (or versions of) Christian supernaturalism. In the form of new texts and ideas, the East was accommodated to these structures
>
> (Said 1985: 122).

Therefore, academics and the people they study "construct stylized images of the Occident and orient in the context of complex social, political, and economic conflicts and relationships ... These stylized images are not inert products. Rather, they have social, political, and economic uses of their own, for they shape people's perceptions, justify policies, and so influence people's actions" (Carrier 1995: 11).

In other words, Europeans look at Asian values with Western eyes, while Asians view Western values with Asian eyes. Being an outsider is partly an advantage, partly a disadvantage, when investigating the values of others. As Levi-Strauss rightly stated, "It is from inside that we can apprehend the ruptures but from outside that some effects of coherence appear" (Levi-Strauss 1966: 125). While the insider has access to the details, the outsider has to rely on limited first-hand experience and secondary sources. However, the horizon can be wider with a more distant view. The task for a researcher is to reveal these distinctive structures of meaning.

An attempt should be made to analyze a culture on the basis of its own "logical" structure. In each culture, one must therefore focus on the so-called *archetypes* rather than on the formal and often officially propagated manifestations of a culture (further elaborated in Servaes 1999). Along with other anthropological research methods, the study of cultural expressions, such as art, folktales, film, video, or literature (see, e.g., Harrison 1994; Harrison and Jackson 2010; Servaes and Malikhao 1989) can assist to understand this "logical" structure.

In other words, in the study of concrete examples of cultural identity, one must be attentive to the following aspects:

(a) the characteristics and dimensions of the cultural reference framework (i.e., the world view, the ethos—norms and values—and their symbolic representation); (b) the interaction and interrelation with the environment of power and interests; and (c) the "ideological apparatus" by which the cultural reference framework is produced and through which it is at the same time disseminated.

Important starting questions are:

- How do Thais construe and interpret their own "Weltanschauung" (world view)?
- How do they explain their world in terms of (wo)mankind, (wo)man to (wo)man relationships, (wo)man to nature, and (wo)man to the supernatural relationships?
- What are the formats, contents, and institutions in which such a world view and value system are symbolically represented?

As the needs and values that various communities develop in divergent situations and environments are not the same, various cultures also manifest varying "identities." Far from being a top-down phenomenon only, foreign mass media and cultural influences interact with local networks in what can be termed a "coerseductive" (for coercion/seduction) way. Far from being passive recipients, audiences are actively involved in the construction of meaning around the communication messages they consume. Consequently, such messages may have different effects and meanings in different cultural settings.

Mainly expanding on the cultural, anthropological, and sociological interpretations of Chamarik (1993), Klausner (1983, 1997), Mulder (1985, 1990, 2000), Phongphit (1989), and Rajadhon (1968, 1987), I attempt to offer an analysis of the interdependency between Thai culture and its communicative expressions with the aid of two complementary, mutually interpretive, and influential dimensions from the traditional, rural, and animistic culture, which still fundamentally condition modern-day Thailand. One dimension is of a spiritual-moral nature, the other is a sociological one.

4.2 Thai Feudalism: The "Sakdina" System

Historically, the Thai societal structure is rooted in the so-called *Sakdina* system. (*Sakdi* means status or power and *na* means land or rice field. Sakdina could therefore be translated as "land status" or "status shown by land.") The major difference between the Sakdina system and the European feudal system is its dependence on the king and the changeability of status. Status was not possible unless one had royal blood. The king or *Chao Phaendin* (the lord of the land) was perceived as infallible, semidivine, and all-powerful. He was the only land owner. He distributed the right to use land according to the Sakdina status which depended in turn on an individual's relationship to blood or by service to the king. The closeness of that relationship had to be ranked with great precision because the Sakdina status determined an individual's rights, wealth, political power, and responsibilities to the state as well as his/her relationship to the rest of society. According to Somsamai Srisootarapan (pseudonym of Jit Pumisak, a famous artist and critical scholar), the major characteristics of the Sakdina system are as follows: (1) The king was the owner of all land, with absolute power over land and people; (2) the people did not have the right to own land. They had to rent the land and pay back with produce at high rates; (3) there was an exploitive relationship between landlord and serf; and (4) the king's officials were given land, horses, buffaloes, and men so they could exploit common men for personal and royal benefits (Srisootarapan 1976: 91–92).

The introduction of capitalist modes of production hasn't fundamentally altered this Sakdina system (Baker and Phongpaichit 2005; Keyes 1989; Terwiel 2005). The Sakdina system modernized materially without changing its psychological dependence on the old traditions of power. Therefore, Wedel and Wedel (1987: 23) state that

> the monetization of the economy eventually forced the old system of land control to become one of private ownership. Land changed from the means of subsistence to just another commodity that could be sold. This change and the failure of the Thai peasant to understand it, at least initially, worked to concentrate land in hands of many fewer people. This created problems of land ownership that persist today.

The natural alliance of the Chinese capitalists and old aristocratic families began to be expressed in convenient marriages that joined economic and political power (Charoensin-o-larn 1988). Therefore, "the transition from a feudalistic to a capitalistic society leaned on rather than destroyed the conservative force … (and) the formation of a public consciousness through State or military-owned mass media has also brought another form of feudalistic thought" (Lertvicha 1987: 59; see also Prasirtsuk 2007). TV series and films which portray or promote the Sakdina values, like *Look Tas* (Slave's child) or *Nang Tas* (Slave wife), have remained very popular.

4.3   A Rural Village Culture

In the Thai rural village culture, the family is the center of the local community. The family (or inner-group) guarantees security and stability. All members of this inner-group know their place and role and act accordingly. Social cohesion and group identification, as well as social control, are very profound. Life is hierarchically orientated on the basis of mutual trust and moral kindness. In Thai, this can be termed as *obe-orm-a-ree* or *pra-khun*. Outside of this protective and "safe" world of the family and the village, the individual enters into a threatening and chaotic "outside" world. The principles of moral kindness, which are highly regarded in the inner-group, are of a far lesser importance in the world of the outer-group. There, the so-called law of the jungle rules, the amoral and sometimes also the immoral worldly and supernatural power (*am-naj li-lab*). To be able to survive personally and socially, the individual has to beg for the protection and the favors of these powers and not upset them in any way.

Power is the most central element in the Thai worldview. The way in which Thai deal with power and submit themselves to it is essentially animistic. Animism does not attempt to explain the complexity of everyday reality on the basis of "rational" and universal principles, but reduces the world into simplistic categories such as "us" and "them," insiders and outsiders. The insiders and their ancestors form the natural center of the universe. The inner-group offers continuity, stability, and protection; outside of that there is chaos and danger in which the Thai feels threatened by all sorts of supernatural powers. In this realm of "phi" (spirits), there are many amoral, but also immoral spirits who each are supposed to be favoured with a certain function or quality. Some are more active than others, but no definite hierarchy can be perceived. So what matters most for the Thai is to try to stay on friendly terms with as many as possible, or to protect against them. The very first word almost every Thai child learns is "phi" and to this word is generally appended the adjective "laug" (*haunting*). They are warned that if they do not behave in a pleasing and appropriate way, they will have to give account to the "phi laug" (haunting spirit), who will mete out the severest punishment. In almost every house, one will find a *phra-phoom* or a spirit house. The house spirit is believed to protect the land and its inhabitants for as long as they honor him in a suitable manner (e.g., by paying homage to him and/or bringing him some food). Some spirits are believed to protect certain places such as forests, mountains, and land; others cure illnesses, assist in protecting from accidents, or "reveal" the winning lottery numbers. At the same time, the Thai can, with the aid of objects, actions, or persons of influence, protect themselves to the possible danger of haunting spirits. Holy objects (sing saksit) such as Buddha images or amulets or pha pra-jiad or pha-yan (a piece of cloth inscribed with mantra), astrologers, magic monks are believed to offer protection. In this way, the Thai surround themselves with a sort of protective "aura" (Srichampa 2014).

Everyone, that is both a person with good or bad intentions, can invoke protection or favors of a higher power or can try to neutralize a potential negative influence in the outer world. To do so they have to follow a clearly laid out ceremony or ritual that can be different from spirit to spirit. For instance, some spirits are believed to like entertainments such as classical dance, others prefer red soda pop or khanom bua-loy (a kind of dessert soup made of glutinous rice balls in sweetened coconut milk). As the artifacts of the society change, the "taste" of the spirits also changes. If a wish is not granted, the Thai dare not question the expertise or power of the spirit. Rather, they will look for a cause or explanation in their own behavior. Perhaps the individual did not honor the spirit in the correct way, or perhaps the spirit's pride was, for unknown reasons, injured. Another often quoted explanation may be that another power stood in the way of the fulfillment of the wish. As it is characteristic of these spirits, because their power does not come from elsewhere, no higher legitimation exists. Therefore, different powers can exercise opposing influences and thwart the fulfillment of a wish.

In other words, to survive in this threatening and chaotic world, a Thai has to make a sort of allegiance with these powers. This allegiance is of a business-like, non-emotional and limited sort. Moral judgements form no part of the contract.

Not opposing, but rather complementary to this amoral power, is the dimension of moral kindness and mutual trust in the inner-group. Everyone is daily confronted with both. Whereas *pra-dej* represents the amoral order and immoral chaos, *pra-khun* represents the moral backbone and stability. Whereas power is mainly symbolized by supernatural projections, one finds *pra-khun* mainly in worldly manifestations. Power is strong, whereas kindness is weak. Power is masculine, whereas kindness is feminine. The main symbol of moral kindness is the mother. The love she has for her children is pure, disinterested, and unconditional. Also elderly people and *ajarn* or "educators" are, because of their unconditional efforts, their knowledge, and morals, regarded as *phu mee pra-khun* (those whom we owe gratitude). Because relations based on kindness are not business-like but moral, they have to be valued in an appropriate way. For example, a lack of respect or gratitude (*a-ka-tan-yu*) is regarded as disgraceful and is punished on the basis of the moral justice principles. Psychologically, this is a source of many guilt feelings.

4.4   The Power of Beliefs

Instructions on the way in which to observe kindness and morality are embodied in Buddhism. The *Dhamma,* the Buddhist teaching, provides the moral instructions by which each individual can surpass the worldly order of passion and *defilement* (*kilesa*), rebirth (*samsara*), and fate (*karma*). If one follows the instructions of the Dhamma, lives according to the eight noble principles, one can reach the highest worth, the *nirvana*, the goodness of pure worth, pure humanity, and pure wisdom. In the *Hinayana* or *Theravada* Buddhism this path has to be walked individually, salvation is no gift from heaven but a permanent, lifelong mission. Sivaraksa (1981:72) summarizes it as follows:

> Buddhism emphasizes the middle way between extremes, a moderation which strikes a balance appropriate to the balance of nature itself. Knowledge must be complete knowledge of nature, in order to be wisdom; otherwise, knowledge is ignorance. Partial knowledge leads to delusion, and encourages the growth of greed and hate. These are the roots of evil that lead to ruin. The remedy is the threefold way of self-knowledge, leading to right speech and action and relations to other people and things (morality), consideration of the inner truth of one's own spirit and of nature (meditation), leading finally to enlightenment or complete knowledge (wisdom). It is an awakening, and a complete awareness of the world.

For more details about Thai Buddhism, I recommend reading Buddhadasa (1986), Hattam (2004), Punyanubhab (1981), Rajavaramuni (1983), Sivaraksa (1988, 2009), or Thipayathasana (2013).

But as, in my opinion, the Thai's common beliefs are more of an animistic than a Buddhist nature, the Thai is more oriented toward the exterior, institutionalized norms of Buddhism (the temple, Buddha images, monks, ceremonies), than toward the moral principles of Buddhism. Supernatural elements such as spirits and ghosts form a substantial part of many videos which are shown on TV, such as *Mitimued* (Dark Dimension), or *Peesad Saen Kon* (A tricky ghost). An example of a communication campaign in which "animistic" symbols have been successfully used is the *Magic Eyes* or *Keep Thailand Clean*-campaign, which was launched in 1984 and organized by the Thai Creative Society. Magic Eyes conforms to the Thai belief that when they do something wrong a spirit will watch them. The Magic Eyes campaign, which ran on all four TV channels and made use of cartoon films, strove to introduce the belief that somebody will watch you whenever you are careless about the environment.

That does not mean that the Thai's animism does not correspond with Buddhism. Rather, it is the other way around. Certain basic principles of Buddhism, such as *Anicca* (everything is not perpetual), *Dhukka* (life is full of suffering), and *Anatta* (the perception of self is illusory) fit in completely with the Thai common belief. Therefore, Niels Mulder (1985:44) concludes that

> the Thai Weltanschauung combines the sophisticated elegance of a universal principle with the primordial directness of animistic thinking; somehow Therevada Buddhism and the Buddhist animistic heritage have corroborated and concluded a perfect marriage. The Buddhist message does not endow this universe with a center to cling to; but characterizes this-worldly and this-cosmic existence as impermanence, suffering, and nonself, guided by the impersonal this-cosmic principle of karma.

In other words, the Thai Buddhist perceives his/her world view as essentially "supernaturalistic." He/she "sees" all phenomena as an integrated whole, in a "sacred" rather than a "secular" world, a cosmos that is to a large extent governed and controlled, not by just the human powers-that-be, but by the supernatural

Table 4.1 Summary of the three religious subsystems of Thai religion

	Buddhism	Brahmanism	Animism
1 Goal orientation	Other worldly	This worldly	This worldly
Worldview	Rational/certain	Rational/certain	Capricious/uncertain
Ritual	Standard/routine	Standard/routine	Tailored to individual
2 Specialists	Mainly male	Mainly male	Mainly female
Recruitment	Universalistic achievement	Universalistic achievement	Particular ascription
3 Participants	Laity	Client	Client
Involvement	Constant	Intermittent	Intermittent
4 Attitude toward	Highly favorable	Favorable	Ambivalent
Social focus	Whole society	Bridging local and society	Highly local

powers. Also Davis (1993:35) argues that Thai religion has three components or subsystems, comprising Therevada Buddhism, Brahmanism, and animism which mutually support each other without conflict. He sets these out in Table 4.1.

4.5   The Amoral Power … and Moral Kindness

The institutionalized way of living of the Thai is set in this continuum with, on the one hand, the moral order and, on the other hand, the amoral power. Relations in the *phra-khun* dimension, inside of the hierarchy of the inner-group, are based mainly on mutual trust and informal dependency. Relations in the *phra-dej* dimension, inside of the hierarchy of the power-based outside world, are characterized by mutual distrust and formality. Because the Thai view of the world lacks a center in which the opposites between power and morality can be overcome, they have to take notice of both in everyday life and have to act according to the situation in which they find themselves.

This does not mean to say that relations of which both dimensions conduce do not exist. In the mass media, the symbol of the "good and rightful" leader, father, village elder, manager, minister, or general, who manages to combine kindness and power inside one person, often appears. However, the outcome of this kind of story is very often rather negative. An explanation for these failures is given neither in the media products nor in reality. In view of the fact that Thai worldview is mainly controlled by supernatural powers, Thais do not have a need to look for "rational" explanations. They accept the world as it is and adopt a rather fatalistic attitude toward social change. In given circumstances, they try to make the best of it for themselves and their immediate relatives and friends. For instance, in the novel "The Judgement" by Kobjitti (1983), which has also been turned into a widely popular film, the moral advice underlying the story can be summarized as follows:

> This is your karma. This is what I spoke with you about. The world outside is in a state of turmoil. At that time you weren't prepared to believe me … Who do you want me to tell? If what you've told me is not true, then—I don't want to get a reputation as a monk who lies. But if it is true, how will I force them to believe me when they've already decided that they don't want to believe you? Just try to do good every time you're in their presence. You'll feel better
>
> (Kobjitti 1983: 38–39).

4.6   Interpersonal Communication: Mai Pen Rai

Some people explain the friendly, modest, and conflict-avoiding way in which the Thai behave in public life as being in accordance with the Buddhist principle of the middle way. To me, however, it seems as if an explanation based on the Thai animism is more appropriate. Contacts with people in the power circles are business-like, formal, and instrumental. To engage oneself for long periods, or to show any feeling is therefore not necessary. But on the other hand, one has to be careful not to upset the power's feelings, and honor it in the proper way. For that reason, children are taught at a very early stage in life to suppress their emotions, and to avoid open conflicts. This behavior, known as *jai yen* (literally: cool-hearted) is a fundamental contrast to the Western more assertive way of communicating. Besides avoiding conflicts (*kreng-jai*), social interactions are characterized by formality, superficiality, and an easy-going atmosphere. This attitude can be best typified by the common saying: *mai pen rai*, never mind, take it easy. Two of the concepts that appear in virtually every conversation, and are a gauge of the way in which the Thai dream about life, are *sanuk* (fun, amusement) and *sabai* (pleasure, comfort). Therefore, the ingredients which form a substantial part of almost all popular Thai audiovisual plots, and which are considered *sanuk* by Thai standards, are violence, romance, superstitious events, and *talok ba-ba-bor-bor* (i.e., a kind of humor which could be defined as "silly" or "slapstick" by Western tastes). Also Western (especially American) film- and TV formats are copied and adapted to Thai tastes. In 2016, the popular Woody Kerd Ma Khui (*Woody born to talk*) and Jao-Jai (*Piercing right to the heart*) talk shows, for instance, imitate popular American talk shows in the sense that the host would interview in a straight-forward and direct/assertive way like an American. All it adds to the American format, for certain programs, is more comedy and silly games. For those interested in a combination of silly games, singing performances, comedy acts, with comedians or stars as talk show hosts, there are *Khor Rong Ya Yud Rong* (*Please do not stop singing*), the *Tee Sip Day* (*At Ten Day*), or *Ching Roi Ching Larn* (*Compete for 100 million*). Also "animistic" series such as *Khon Aoud Phi* (*Challenging Spirits*), *Mu Night* (*Black Magic Night*), *and Scan Karm* (*Scan your karma*) are still popular in 2016. Prasertkul (1989a: 64), one of the student leaders in the 1970s, observed that "whenever Thais meet, they try to make others laugh even though the subjects they are talking about may not be relevant for jokes. Life talk shows are very popular, especially during election campaigns... (also) Thai newspapers have a special expertise in transforming news into entertainment."

In interpersonal communication, the end product, the content, tends often to be overlooked. Not many take notice of this as in the Thai society, the social achievement, the form, the show element are that much more important. Therefore, false modesty has no place in Thailand. Outward characteristics of status and power are fully and emphatically shown. Despite the superficiality and "showing-off" the power expects to be recognized and respected in a suitable fashion. Besides, power is amoral, and value-verdicts have no part in the power game.

It is often difficult for an outsider to distinguish between the different, moral, or amoral relation patterns. The outward signs are often the same. But for the Thai, the subtleties and nuances are immediately clear. The differences appear in the use of language as well, not only in the way of addressing, but also in the description of the different behavior patterns. To show respect for elders or teachers is described as *krengjai* (to respect somebody with the heart), and to render honor to a representative of power on the other hand is called *krengklua* (to respect out of fear). The meaning behind certain actions can differ completely as well. To return gratitude (*tob thaen bun-khun*) in an inner-group relation is based on the principles of morality and commitment; in an outer-group relation this is, however, of a formal and business-like character. Thais do not approach prostitution, for example, from a

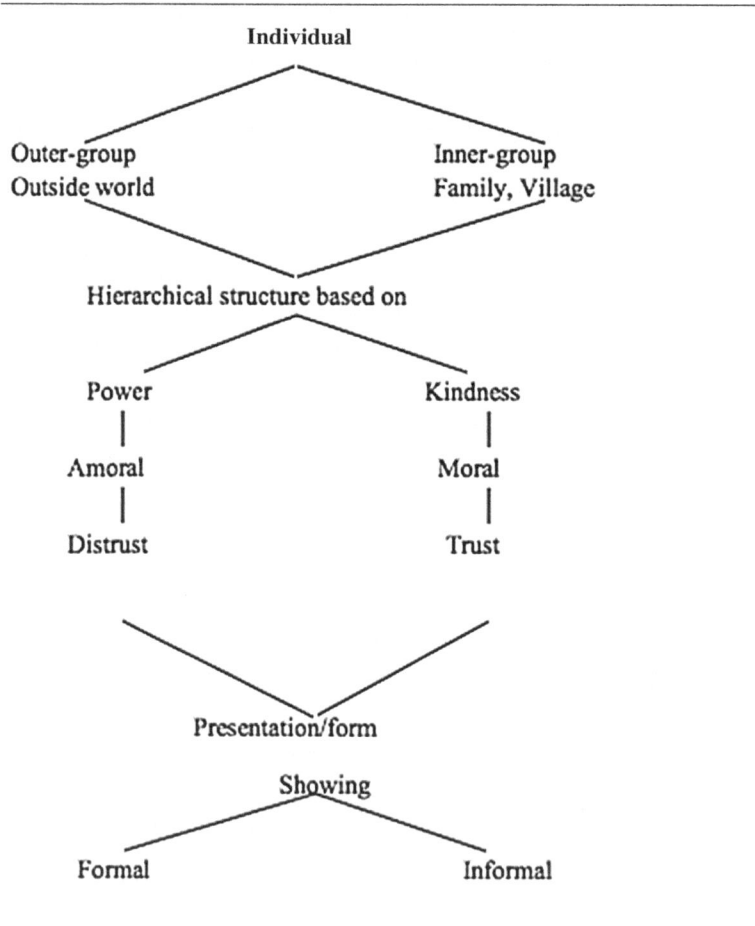

Fig. 4.1 Main characteristics of Thai culture

moral criterion, but on the basis of relations within the power sphere (Jeffrey 2002; Meyer 1988). More of these everyday situations, which are often said to be "incomprehensible" to the Western world, are described in Holmes and Tangtongtavy (1995), Klausner (1997), Knutson et al. (2012), and Komolsevin et al. (2011).

In Fig. 4.1 I have attempted to sum up the most important characteristics of the Thai culture:

4.7 Likay Drama

Buddhists and animists stress the progression and the temporality of life. Having fun cannot last long enough. Therefore, most Thai movies easily last for two to three hours. They have adopted the format of the *likay-* drama play, which is still very popular in rural areas. Likay is a traditional theater form or folk opera of Malay or Indian origin, thought to have been introduced to Thailand toward the end of the nineteenth century. It combines music, singing, drama, and narration, with the performers dressed in traditional Thai costume. Its plots are based upon those of folk legends. Often the story will be serious, with the costumes identifying the villain, the victim, and the hero. Spectators get so caught up in the dramatic presentation that they often weep and may even threaten the villain from the audience. Levity is also included, usually by one or two "simple-minded" characters. (For more information, see Virulrak 1975; Barry 2013.) In likay, the message and the medium are closely tied. Therefore, Davis (1993: 114) argues that "to effectively communicate the message it needs to be packaged in a medium that will be culturally acceptable. For Thai this medium is drama. There seems to be no other means of communication which makes such a powerful impact on Thai audiences as the dramatic arts."

4.8 Growing Pains: Modernization and Tradition

Modernization and Westernization have strengthened the animistic concept of power rather than weakened it. Unemployment, economic, and political crises have made life for the majority of the population even less attractive, so the need for worldly and spiritual "protection saints" has increased proportionally, just as the struggle for status and prestige under the growing Westernized middle classes has increased. Whereas before one fought with traditional "power-means," the power of money is undermining many traditional relationship patterns. The impersonal and uncontrollable money is ever present and even throws a shadow on certain *bun-khun* relationship patterns. So it is not surprising that corruption and the misuse of power are on the rise:

> The prevalence of bureaucratic corruption stems from the systems of self-remuneration in the traditional bureaucracy. Officials were expected to remunerate themselves by taking a cut from revenues they collected, and extracting fees for services performed. In the transition to a modern form of bureaucracy, these practices were never erased. Meanwhile the systems for imposing moral and conventional limits on the extent of such self-remuneration have tended to decay… The rise of corruption as an issue was more a function of increasing

competition for political power and corruption revenues between the old power-holders in the military and civilian bureaucracy, and the new challengers in civilian politics, particularly those with a business background

(Phongpaichit and Priryarangsan 1994: 173).

In the hierarchically structured Thai society in which form and performance play a major role, the individual is regularly confronted with situations that cause heavy psycho-social stress. As Thai rationalize these psycho-social problems in an animistic and fatalistic way, as the work of a bad spirit and so on, they therefore do not seem to be able to address these tensions. The only way to solve such a problem is, in their opinion, to get rid of the bad spirit by eliminating it (or its personification) combined with the propensity to suppress anger, frustrations, and so on, until it literally bursts. Whenever this happens, the outcome is usually very extreme and this, for instance, is one of the explanations for the fact that Thailand has one of the highest crime rates in the world. "With murder rate equal to US and three times Cambodia, Thailand's 6.1 million registered firearms used in never-ending stream of murder cases with loss-of-face and business disputes two main motives," reports Jon Fornquest in the Bangkok Post of August 11, 2015.

As more people fall by the wayside in this power struggle, the longing for a "safe" life in the inner-group increases. On the political front, this tendency has given rise to the revival of conservative and nationalistic ideas. On the personal level, it leads to the strengthening of Brahmanical and spiritual practices. The result of a survey on Thai values among urban and rural Thai indicated that certain superstitious behaviors such as "fortune-telling" and "lucky numbers" are practiced more among Bangkokians than among farmers. No difference was observed in terms of educational level. "This casts some doubt on the theory that postulates a negative correlation between education and supernatural belief and behavior. However, it is a dominant value behavior characteristic of the Thai. In addition, it is a known fact that a number of highly powerful people in Thailand have their personal well known fortune-teller" (Komin 1988: 171). As many politicians and business people before him, the first thing military coup leader General Prayuth did after toppling a democratically elected government on May 22, 2014, was to go through the rituals of animism and spiritualism to ward of all kinds of evil. "Despite its outwardly modern appearance, everyday life in Thailand still prominently features pre-Buddhist animist beliefs," observed Amy Sawitta Lefevre (2014).

Public life is organized on the basis of friendship circles with an influential leader on the top, that is the so-called *patronage system* (Chaloemtiarana 1983; Lertrattanavisut 2004; Nelson 2004; Terwiel 1984, 2005). Thais do not follow political programmes or abstract ideas but follow leaders and charismatic figures on the basis of the 'right or wrong, my group'-principle. Girling (1981, 1984), who applied the Gramscian hegemony principle to the Thai society, reports that the production basis is integrated in and determined by the culture-ideological superstructure of the civil society: "The result, in Thai terms, is the 'bureaucratic policy', or what Gramsci calls 'transformism': a ruling class that grows ever more extensive by absorbing elements from other social groups who then operate within the established framework" (Girling 1984: 445). He contends that in these circumstances there is little chance for social change. From a culturalistic perspective this view is confirmed by Rajadhon (1968: 29): "The social system, habits and customs as seen in modern times are superficial modifications of the fundamentals and in a comparative degree only."

4.9 The Thai Value System

Generally speaking, the Thai social system is essentially a society where "self-centeredness" and interpersonal relationships are of utmost importance. Even though the Thai self-image is often described as individualistic, I prefer to term it as a "weak" rather than a strong personality. This is also the opinion of Brummelhuis and Kemp (1984): "The individual's preoccupation is not so much with self-realization and autonomy as with the adaptation to the social or cosmological environment. If a notion of Thai individualism is to have any specific meaning it is in designating that particular mode of retreat, avoidance and distrust, which colors so many forms of behavior and social relationships." (Brummelhuis and Kemp 1984: 44–45; see also Kashima et al. 2002). Prasertkul (1989a: 64) is more critical: "Our national traits, which I think are very strong, are: firstly, Thais do not like serious matters; they like to crack jokes and talk about sensational matters, especially dirty 'under the belt'-matters. Secondly, they are egotist. They use group benefits to be their norms. If matters are not relevant to their own lives, they will not take them into account." That kind of egoism leads Mulder (2000: 49) to conclude that "it is not unlikely that Thai society generates culturally unique psychological tensions by repressing individuals through identifying them with their presentation, thus stifling their needs for self-expression and open communication." For instance, commenting on the annual deadly road accidents, the third highest in the world according to the World Health Organizations (WHO) and the almost nonexistent follow-ups by police or policy makers to address these issues, Techawongtham (2016) explains:

> In my opinion, this 'attitude' is part of a larger and more troubling Thai character trait embedded in our cultural DNA. The Thai word for this is *mak ngai*. It may be translated roughly to English as the predisposition to take an easy way out without regard to consequences or others (and often to oneself as well). Simply put, it's a 'me first and me only' attitude. We can see this attitude expressed everywhere in everyday life.

One of the scholars who has been trying to find value patterns in different cultures is Hofstede (1980, 1991). He surveyed over a hundred thousand workers in multinational organizations in more than fifty countries and identified a number of value dimensions that are influenced and modified by culture. In their latest work (Hofstede and Hofstede 2005), they identified six value dimensions that are influenced and modified by culture: (a) individualism-collectivism, (b) uncertainty avoidance, (c) power distance, (d) masculinity and femininity, (e) long- or short-term orientation, and (f) an activity orientation. This theory of cultural variability attempts to assess the range in which countries differ in cultural values on a continuum.

Referring to the above Thai Value Study, Komin identified nine value clusters according to their significant positions in the Thai value system, namely:

(1) ego-orientation (which is the root value underlying various other key values, such as "face-saving" and "*kreng-jai*,"),
(2) grateful relationship orientation ("*tob thaen bun-khun*," or paying back in kind or "*mee kwam katanyoo*" (the realization that one must pay back gratitude.),
(3) social smoothing relationship orientation (caring, pleasant, polite),
(4) flexibility and adjustment orientation (situation-orientedness),
(5) religio-psychical orientation (believe in *karma*, having superstitious beliefs),
(6) education and competence orientation (form is more important than substance),
(7) interdependence orientation (peaceful coexistence of ethnic, religious, etc., groups),
(8) achievement-task orientation (achievement is the least important value among Thai, and it connotes social rather than task achievement), and
(9) fun-pleasure orientation (fun loving is both a means and an end in itself). "These are the major value orientations registered in the cognitive world of the Thai, and serve as criteria for guiding behavior, or as the blueprint that helps to make decisions at the behavioral levels" (Komin 1988: 172).

She explains that these value orientations have to be taken into consideration in any development program as they often prove to be "stumbling blocks" to social change. One of the big stumbling blocks for Thais in the future ASEAN Economic Community (AEC) is the education system and the lack of English proficiency. Being able to speak English is an important criterium for a Thai to get a good job.

However, the Bangkok Post (2015), in a special issue on the state of Thai education, observes "We No Speak English. Language teaching and learning are in a parlous state in the Kingdom," which is further commented by Echachai (2015): "The goal of our education system is to instill youngsters with a set of cultural values that endorses and perpetuates the existing social and political hierarchy... Our education system has utterly failed to help our kids be competitive in the modern world because it suppresses creative and independent thinking" (see also Scientist 2015). In addition, Buripakdi (2014: 103) reports in her research on the ability to use English by Thais that people are selected on the basis of their ability to produce an accent that is close to that of a native English speaker rather than assessing the content of what that person can deliver.

In Table 4.2, I summarize the Thai system of worldviews and values and provide a few examples of the resulting symbolic representations.

More examples of these symbolic representations in interpersonal and mass communication in Thailand are (see also Malikhao 2012, 2014; McCargo 2000; Servaes et al. 2009; Seypratub 1995):

Table 4.2 Thai worldviews, values, and symbolic representations

Worldview	Value system	Symbolic representations
Brahmanism and its interplay with Buddhism and animism	Hierarchy	• Language use • Gender inequity regarding sexual norms • Social class • Seniority
Animism and its interplay with Buddhism and animism	Power	• King • Military • Amulet industry • Spirit houses • Chao por/chao mae and mafiosi • Politicians
Theravada/Hinayana Buddhism and its interplay with Brahmanism and animism	Interdependence orientation	Peaceful coexistence of ethnic groups
	Social smoothing relationship	• Smiling (jai yen) • Mai pen rai (never mind) • Kreng-jai • Modesty • Conflict avoiding • Face-saving relationship
Hinayana Buddhism (emphasis on individual enlightenment experience)	Social achievement orientation	• Indulgence in titles and ranks • Honorary degrees • Decorations
Village culture (family-centered through moral kindness and mutual trust)	Grateful relationship orientation	• Dek sen/dek phak (nepotism) • Child labor • Sex work • Voting
	Fun-pleasure orientation	• Dancing in demonstrations • Family planning • Gambling • Amusements during funerals

- Hierarchy

    - Royal news comes first on TV and radio,
    - Layout of newspapers—royals on the top,
    - Elite-centered columns, e.g., sappy Hi-so, and
    - Girly magazines/sexy girls on front page of Sunday newspapers.

- Power

  - Military-controlled TV and radio channels,
  - Animistic magazines,
  - Astrologers' columns,
  - Monks (as astrologers) and astrologers,
  - Grateful relationships, and
  - Sexual relationships between starlets and rich business people, media producers, and media owners.

- Grateful relationships

  - Bribery, kickback, and tea money (such as paying DJ's and reviewers).

- Social smoothing relationships

  - Gossip and rumors and
  - Crime and violence.

- Interdependence orientation

  - Chinese and English newspapers allowed.

- Social achievement orientation

  - Khun ying news and
  - Donation lists (on radio, TV, temple walls/window frames, etc.).

- Ego-orientation

  - Mercedes and cell phones and
  - Media are manipulated by those in power.

- Fun-pleasure orientation

  - Human interest news,
  - Star gossips on front page of newspapers,
  - Jokes in newspaper, and
  - Dirty jokes.

4.10   Thai Leaders and Face

As power is essential in the Thai hierarchical culture, I wish to briefly elaborate on the role and place of Thai leaders in this context, using the new and interesting work of Larry Persons (2008, 2016) as a guide. Persons studied the different types of leaders with their different faces of power. He joins Mulder (2000) when he argued that "power is the central axis around which public life revolves" (Mulder 2000: 140). However, he prefers to use the related concept of "face": "Thai leaders use face as a resource for their personal empowerment in leadership" (Persons 2016: 134).

He then borrows the three "leadership foundations" which emerged from the ethnographic fieldwork of Conner (1996), who found three sources of power available to Thai leaders: *amnaj*, which he calls "authority"; *itthiphon* or "influence"; and *barami* or "personal power." While *amnaj* is based on legal or institutional power, *itthiphon* comes from a leader's access to and control of certain valuable resources that others need, and barami refers to the moral strength and "selfless moral behavior" of leaders. While *barami* could be placed in the *phra-khun* dimension, inside of the hierarchy of the inner-group, based mainly on mutual trust and informal dependency; amnat and especially itthiphon are power relations in the *phra-dej* dimension, inside of the hierarchy of the power-based outside world, characterized by mutual distrust and formality (see Fig. 4.1). However, "*itthiphon* is by far the most pervasive and effective form of power in Thailand today. To wield this kind of power you must accumulate and control coveted resources, things that other people dearly need, or at the very least, greatly desire" (Persons 2016: 150). This kind of power can be rooted in socioeconomic force (money), physical force (ability to gain compliance by physical coercion), or psychological force (the threat of using one or both of the former). So "*itthiphon* leaders have two primary tactics for controlling their entourages. They woo them with kindness and promises of profit, and they restrain them with displays of power. In other words, these leaders want subordinates to both love them and fear them at the same time" (Persons 2016: 153).

While the three forms of power/leadership continuously overlap and interact, and while people, from a traditional village perspective, may still long for the barami in their leaders, they most often are faced with the hard reality of itthiphon instead.

4.11 Conclusion

1. An analysis of the interdependency between Thai culture and its communicative expressions was offered with the aid of two complementary, mutually interpretive and influential dimensions from the traditional, rural, and animistic culture, which still fundamentally condition modern-day Thailand. One dimension is of a spiritual-moral nature, the other is sociological.
2. Power is the most central element in the Thai worldview. The way in which Thai deal with power and submit themselves to it is essentially animistic. Animism does not attempt to explain the complexity of everyday reality on the basis of "rational" and universal principles but reduces the world into simplistic categories such as "us" and "them," insiders and outsiders.
3. The institutionalized way of living of the Thai is set in a continuum with, on the one hand, the moral order and, on the other hand, the amoral power. Relations in the khuna dimension, inside of the hierarchy of the inner-group, are based mainly on mutual trust and informal dependency. Relations in the *pra-dej* dimension, inside of the hierarchy of the power-based outside world, are

characterized by mutual distrust and formality. Because the Thai view of the world lacks a center in which the opposites between power and morality can be overcome, they have to take notice of both in everyday life and have to act according to the situation in which they find themselves.

4. In Thailand, as documented in other parts of the world as well, one observes at least two interrelated developments with regard to the production and consumption of media software and hardware. On the one hand, there is a tendency to import cultural content and technology and develop local imitations, that is, attempting to forge a more autonomous culture, independent of but at the same time borrowing from foreign (mainly Western) cultures. On the other hand, as is the case in the West, one observes that in spite of the better production quality the majority of local audiences prefer programmes produced in their own culture.

5. Modernization and Westernization have strengthened the animistic concept of power rather than weakened it. Unemployment, economic, and political crises have made life for the majority of the population even less attractive, so the need for worldly and spiritual "protection saints" has increased proportionally, just as the struggle for status and prestige under the growing Westernized middle classes has increased. Whereas before one fought with traditional "power-means," the power of money and *itthiphon* is undermining many traditional relationship patterns.

References

Baker, C., & Phongpaichit, P. (2005). *A history of Thailand*. Melbourne: Cambridge Press.

Bangkok Post. (2015). We no speak english. Language teaching and learning are in a parlous state in the Kingdom. What is to be done? In *International Education 2015–2016*. Special Publication. Bangkok Post, Bangkok.

Barry, C. (Ed.). (2013). *Rights to culture. Heritage, language, and community in Thailand*. Chiang Mai: Silkworm.

Brummelhuis, H. T., & Kemp, J. (Eds.). (1984). *Strategies and structures in Thai society*. Amsterdam: Antropologisch-Sociologisch Centrum.

Buddhadasa, B. (1986). *Dhammic socialism*. Bangkok: Thai Interreligious Commission for Development.

Buripakdi, A. (2014). Hegemonic English, Standard Thai, and Narratives of the Subaltern in Thailand. In P. Liamputtong (Ed.), *Contemporary socio-cultural and political perspectives in Thailand* (pp. 95–109). Budoora, Australia: Springer.

Buruma, I. (1989). *God's dust: A modern Asian journey*. New York: Farrar, Straus, Giroux.

Carrier, J. G. (1995). *Occidentalism: Images of the West*. Oxford: ClarendonPress.

Chaloemtiarana, T. (1983). *Kanmang, rabop phokhun uppatham bp phadetkan (Thailand, the politics of despotic paternalism)*. Bangkok: Thammasat University. (in Thai).

Chamarik, S. (1993). *Democracy and development. A cultural perspective*. Bangkok: Local Development Institute.

Charoensin-O-Larn, C. (1988). *Understanding postwar reformism in Thailand*. Bangkok: Editions Duang Kamol.

Conner, D. W. (1996). *Personal power, authority, and influence: Cultural foundations for leadership formation in Northeast Thailand and implications for adult leadership training.* PhD dissertation. Northern Illinois University.

Davis, J. (1993). *Poles apart? Contextualising the gospel.* Bangkok: Kanok Bannasan.

Echachai S. (2015). Education ills only part of our problem. *Bangkok Post*, 2 December 2015 http://www.bangkokpost.com/print/782325/. Accessed December 4, 2015.

Girling, J. (1981). *Thailand: Society and politics.* New York: Cornell University Press.

Girling, J. (1984). Hegemony and domination in Third World countries: A case study of Thailand. *Alternatives, 10* (Winter).

Harrison, R. (1994). Introduction: Sidaoru'ang and the radical tradition in contemporary Thai fiction. In SIDAORU'ANG, *A drop of glass and other stories.* Bangkok: Editions Duang Kamol.

Harrison, R., & Jackson, P. (Eds.). (2010). *The ambiguous allure of the West. Traces of the Colonial in Thailand.* Hong Kong University Press.

Hattam, R. (2004). *Awakening-struggle: Towards a buddhist critical social theory.* Flaxton Qld: Post Pressed.

Hofstede, G. (1980). *Culture's consequences: International differences in work-related values.* Beverly Hills: Sage.

Hofstede, G. (1991). *Cultures and organizations. Software of the mind. Intercultural cooperation and its importance for survival.* London: McGraw-Hill.

Hofstede, G., & Hofstede, G. J. (2005). *Cultures and organizations. Software of the mind.* London: McGraw Hill.

Holmes, H., & Tangtongtavy, S. (1995). *Working with the Thais.* Bangkok: White Lotus.

Jeffrey, L. A. (2002). *Sex and borders. Gender, national identity, and prostitution policy in Thailand.* Chiang Mai: Silkworm Books.

Kashima, Y., Foddy, M., & Platow, M. (Eds.). (2002). *Self and identity. Personal, social and symbolic.* Mahwah NJ: LEA.

Keyes, C. (1989). *Thailand. Buddhist Kingdom as Modern Nation-State.* Bangkok: Editions Duang Kamol.

Klausner, W. (1983). *Reflections on Thai culture.* Bangkok: Siam Society.

Klausner, W. (1997). *Thai culture in transition.* Bangkok: The Siam Society.

Knutson, T., Datthuyawat, P., & Komolsevin, R. (2012). *Teaching in Thailand. A practical guide for expat teachers and trainers.* Wang Aksorn Press: Bangkok.

Kobjitti, C. (1983). *The judgement (Kham Phi Phaksa).* Bangkok: DK Books.

Komin, S. (1988). Thai value system and its implication for development in Thailand. In D. Sinha & H. Kao (Eds.), *Social values and development. Asian perspectives.* Sage: New Delhi.

Komolsevin, R., Knutson, T., Datthuyawat, P., & Tanchaisak, K. (2011). *Thai and American. Communication behavior: A comparison of elephants and eagles.* Bangkok: Wang Aksorn Press.

Lefevre, A. S. (2014). New Thai PM uses holy water, feng shui to ward off occult. http://www. reuters.com/article/us-thailand-politics-blackmagic-idUSKBN0H327T20140908. Accessed 15 May, 2016.

Lertrattanavisut, P. (2004). *Toxinomics.* Bangkok: Open Books Publishing.

Lertvicha, P. (1987). Political forces in Thailand. *Asian Review, 1.*

Levi-Strauss, C. (1966). Anthropology: Its achievements and future. *Current Anthropology, 7*(2), 124–127.

Malikhao, P. (2012). *Sex in the village. Culture, religion and HIV/AIDS in Thailand.* Penang-Chiang Mai: Southbound & Silkworm Publishers.

Malikhao, P. (2014). Thai buddhism, the mass media and culture change in Thailand. *Journal of the Asian Research Center for Religion and Social Communication, 12*(2), 124–143.

McCargo, D. (2000). *Politics and the press in Thailand: Media machinations.* London: Routledge.

Meyer, W. (1988). *Beyond the mask. Toward a transdisciplinary approach of selected social problems related to the evolution and context of international tourism in Thailand.* Saarbrucken: Breitenbach Publishers.

Mulder, N. (1985). *Everyday life in Thailand. An interpretation.* Bangkok: DK Books.

Mulder, N. (1990). *Inside Thai society. An interpretation of everyday life.* Bangkok: DK Books.

Mulder, N. (2000). *Inside Thai society: Religion, everyday life, change.* Chiang Mai: Silkworm Books.

Nelson, M. (Ed.). (2004). *Thai politics: Global and local perspectives.* Nonthaburi: King Prajadhipok's Institute.

Persons, L. S. (2008). The anatomy of Thai face. *Journal of Humanities, 11,* 53–75.

Persons, L. (2016). *The way Thais lead. Face as social capital.* Chiang Mai: Silkworm.

Phongpaichit, P., & Priryarangsan, S. (1994). *Corruption and democracy in Thailand.* Bangkok: Faculty of Economics, Chulalongkorn University.

Phongphit, S. (1989). *Development paradigm. Strategy, activities and reflection.* Bangkok: Thai Institute for Rural Development.

Prasertkul, S. (1989a). Samee Jiab lae Thai society (Samee Jiab and Thai society). *Management Review,* 24–30 (in Thai).

Prasirtsuk, K. (2007). *From political reform and economic crisis to Coup d'Etat: The twists and turns of thai political economy, 1997–2006.* Bangkok: Thammasat University.

Punyanubhah, S. (Ed.). (1981). *Buddhism in Thai life.* Bangkok: Samnakngan Kana Kammakan Raksa Ekalak Thai.

Rajadhon, A. (1968). *Essays on Thai folklore.* Bangkok: DK Books.

Rajadhon, A. (1987). *Some traditions of the Thai.* Bangkok: Sathirakoses Nagapradipa Foundation.

Rajavaramuni, P. (1983). *Social dimension of Buddhism in contemporary Thailand.* Bangkok: Thai Khadi Suksa.

Said, E. (1985). *Orientalism.* Harmondsworth: Penguin Books.

Scientist, A. (2015). Civil discourse and civil society: The dysfunctional culture of Thai academia. *Kyoto Review of Southeast Asia* (Issue 19), July 2015. http://kyotoreview.org/yav/dysfunctional-thai-academia/. Accessed April 20, 2016.

Servaes, J. (1999). *Communication for development: One world, multiple cultures.* Creskill, NJ: Hampton.

Servaes, J., & Malikhao, P. (1989). How 'culture' affects films and video in Thailand. *Media Development, 36,* 4.

Servaes, J., Malikhao, P., & Pinprayong, T. (2009). Communication rights are human rights. A case study of Thailand's Media. In A. Dakroury, M. Eid & Y. Kamalipour (Eds.), *The right to communicate: Historical hopes, global debates, and future premises* (pp. 227–254). Dubuque: Kendall Hunt Publishers.

Seypratub, S. (1995). *Sue muan chon lae karn pattana prathet: Nen chao pao prathet Thai (Mass media and development: The case of Thailand).* Bangkok: Chulalongkorn Publishing House.

Sivaraksa, S. (1981). *A buddhist vision for renewing society.* Bangkok: Thai Wattana Panich.

Sivaraksa, S. (1988). *A socially engaged buddhism.* Bangkok: Thai Inter-Religious Commission for Development.

Sivaraksa, S. (2009). *The wisdom of sustainability. Buddhist economics for the 21st century.* Chiang Mai: Silkworm Books.

Srichampa, S. (2014). Thai amulets: Symbol of the practice of multi-faiths and cultures. In P. Liamputtong (Ed.), *Contemporary socio-cultural and political perspectives in Thailand* (pp. 49–64). Budoora, Australia: Springer.

Srisootarapan, S. (1976). *Chom Nar Sakdina Thai (The face of Thai Sakdina).* Bangkok: Agsorn Sampan. (in Thai).

Techawongtham, W. (2016). Deadly consequences of doing things the 'Thai way'. *Bangkok Post.* 22 April 2016. http://www.bangkokpost.com/opinion/opinion/942797/deadly-consequences-of-doing-things-the-thai-way. Accessed April 22, 2016.

Terwiel, B. (1984). Formal structure and informal rules: An historical perspective on hierarchy, bondage and patron-client relationship. In H. T. Brummelhuis & J. Kemp (Eds.), *Strategies and structures in Thai society.* Amsterdam: Anthropologisch-Sociologisch Centrum.

Terwiel, B. (2005). *Thailand's political history: From the fall of Ayuthaya in 1767 to recent times.* Bangkok: River Books.

Thipayathasana, N. P. (2013). *The natural truth of Buddhism.* Nakornratchasima: Arsharamata.

Van Beek, S. (Ed.). (1983). *Kukrit Pramoj: His wit and wisdom, writings, speeches and interviews.* Duang Kamon.

Virulrak, S. (1975). *Like, traditional folk media of central Thailand.* Paper Seminar on Traditional Media, East-West Center, Honolulu, July–August.

Wedel, Y., & Wedel, P. (1987). *Radical thought, Thai mind: The development of revolutionary ideas in Thailand.* Bangkok: Assumption Business Administration College.

Chapter 5
Tourism, Digital Social Communication and Development Discourse: A Case Study on Chinese Tourists in Thailand

Abstract This paper explains tourism and cultural values amidst the globalization period. It also discusses tourism for sustainability discourse and what sustainable development means from a Thai Buddhist perspective. Since the tourism industry is booming in Thailand, the influx of Chinese tourists has contributed to the Thai economy. However, the cultural misunderstandings between Chinese tourists and Thais have caused a lot of tensions. Via digital communication platforms—such as You Tube, "Kon Dang Nang Clear," "Tang Kon Tang Kid" from Amarin TV discussions about Chinese tourists, and Face Book, of which the parody is "We love Chinese Tourists"—Thais discuss their priorities: money versus cultural preservation as a means of development.

5.1 Introduction

From the official history of Thai tourism in Thailand (TAT 1995), tourism in Thailand started when the first railroad system was introduced to Siam, the former name of Thailand, in 1926. The tourism industry has become known internationally from the mid 1960s when Thailand was the hub for rest and recreation (R&R) for the American GIs fighting in Vietnam. It is undeniable that Thailand has over the past decades become dependent on income from the tourism industry. Millions of visitors roam the country each year to experience what Thailand has to offer: beautiful beaches, mountainous areas, river basins, delicious food, friendly people, and exotic lifestyles. As the cost of living in Thailand is relatively cheap, compared with that of Western countries, many tourists are drawn, not only from Western countries and Japan but also increasingly from China in recent years. Of the 26.5 million tourists who visited Thailand in 2013, 6.3 million came from Europe, 1.3 million from South Asia, and 1.2 million from the Americas. The majority (16 million) came from East Asia. Among these East Asian tourists, 4.7 million were Chinese, the others were Malay (2.9 million), Japanese (1.5 million), and Indians (1 million) (Grossman 2015: 122). As the purchasing power of Mainland Chinese has become stronger, it is fair to say that Chinese tourism has boosted up

© Springer Nature Singapore Pte Ltd. 2017
P. Malikhao, *Culture and Communication in Thailand*, Communication, Culture and Change in Asia 3, DOI 10.1007/978-981-10-4125-9_5

the Thai economy a great deal. In 2015 alone, 4.62 million people came in tour groups and spent almost 188 billion Baht or 5.8 billion US dollars (China daily 2015).

Due to globalization, with its impacts on global aviation and telecommunication, Thailand can be easily accessed and searched as a tourist destination. The Tourism Authority of Thailand (TAT 1995) reports four factors that facilitate the tourism industry: (1) its strategic aviation center in Southeast Asia; (2) its quantity and quality of natural resources and historical destinations; (3) its appropriately developed accommodation, transportation, restaurants, souvenirs, and entertainment; and (4) the friendliness and service-mindedness of Thais. From what TAT explains, globalization and modernization has brought Thailand to the attention of tourists from around the world.

Globalization did not only boost up the tourism industry, but also brought some concerns about cultural heritage, cultural erosion, cultural values, and environmental conservation, among other things. According to Lee (2003: 21–23), globalization has three dimensions: spatial, temporal, and cognitive. The spatial dimension concerns the growing sense of unlimited physical boundaries of social interaction due to advanced telecommunication, speedy travel, trade, and so forth. The temporal dimension concerns the real time and the perceived time in which social interaction takes place. Lastly, the cognitive dimension concerns the creation and exchange of knowledge, ideas, norms, beliefs, cultural identities, and other thought processes. The last aspect of globalization is what this paper aims to discuss. Basically, it is about cultural relativity, stereotyping, and ethnocentrism triggered by tourism.

Former Chinese Premier Wen Jiabao's speech at the British Royal Society, entitled "The path to China's future" (27 June 2011), states as follows:

> We have always called for respecting the diversity of civilizations and advocated dialogue, exchanges and cooperation among them. The late Mr. Fei Xiaotong, a well-known Chinese sociologist, received his PhD at the London School of Economics and Political Science in the 1930 s. Having gone through many vicissitudes in life, he concluded in his late years that 'The world will be a harmonious place if people appreciate their own beauty and that of others, and work together to create beauty in the world.' These thoughts best illustrate the open and inclusive mindset of China today (Xinhua 2014).

From the aforementioned speech, China acknowledges cultural diversity and cross-cultural communication as part of its development scheme. Former Premier Wen refers to democracy, fairness, rules of law, and human rights that would go hand in hand with economic prosperity and sustainable development. It is without a doubt that democracy, human rights, and so forth in the worldview of Socialist China would be differently defined and applied from those in the West. China's open-door policy, as part of joining the globalization process, has made China prosper in economic terms. However, the sociocultural and environmental aspects of the path to sustainable development are still in progress. With economic progress, the Mainlanders can afford more to compete with Western and Japanese travelers to appreciate "others' beauty." However, the quality of the Chinese tourists visiting Thailand in 2015 is somewhat contradictory to what Former

Premier Wen advocated in 2011, regarding respecting diversity of civilization and opting to have dialogue exchanges. At the same time, there has been controversy over the trading off of cultural values, or what the TAT calls "preserving Thainess." Here below are some examples of not so pleasant news on Chinese tourists in Thailand:

- Chinese tourists were banned for half a day from visiting Wat Rong Khun in Chiang Rai after complaints of inappropriate toilet use (Bangkok Post online 2015a).
- A young girl from China was pictured urinating in front of the Anata Samkhom Throne Hall in Bangkok (Shianghaiist.com. 2015).
- A Chinese passenger on Thai AirAsia flight scalded a flight attendant with a cup of hot noodle soup on a charter flight from Bangkok to Nanjing in China because she was not allowed to sit with her husband. Many passengers took pictures of the flight attendant who was screaming in pain (Dailymail online 2015a).
- Police in Chiang Mai, Thailand, tried to locate a presumed Chinese tourist who was filmed kicking[1] a prayer bell at a sacred Buddhist temple (Dailymail online 2015b).
- Up to 500 Chinese tourists a day visit Chiang Mai University campus and disrupt the running of the university.[2] The tourists pitched a tent near the Ang Kaew lake and wrote "we are here" in paint on the ground, caused car accidents, sneaked into classrooms to take photos of teachers and students, and messed the canteen. They even buy or rent a university uniform and pose for a picture. Moreover, on many occasions, the tourists in university uniforms sneaked into classrooms and attended classes (Bangkok Post online 2015b).

Discussion online and negative news about Chinese tourists have raised concerns for Thais to think about ecotourism, cultural relativity, Thai–Chinese relations, and sustainability. This paper aims to study online media discussion on the aforementioned discourses based on the Chinese-dominant Thai tourism industry.

Therefore, the main objectives of the paper are threefold: first, to understand the discourse of tourism for sustainability; second, to explore intercultural communication between Thai-Chinese and to seek common grounds in sustainable tourism via online communication; and third, to study development discourse in lieu of tourism via online social communication.

---

[1]Pointing with feet or using feet instead of hands to touch an object is considered very rude in Thai culture.

[2]These tourists are obviously 'inspired' by the success of a blockbuster comedy movie, "Lost in Thailand", that portrays fascinating landscapes of the Northern region of Chiang Rai and Chiang Mai, and the fun parts of Thai culture, such as eating, drinking, hanging out, and outdoor adventure. One of the scenes unfolds on the Chiang Mai University campus.

5.2 Literature Review and Theoretical Perspectives

5.2.1 Globalization and Culture

Globalization is multi-dimensional and, therefore, requires diverse definitions. Many scholars state clearly that the globalization process in fact has been an ongoing process since the archaic time. According to Hopkins' (2002: 1–10) explanation, the globalization process is a historical process that started before the 1500s. The "archaic globalization" period occurred from Byzantium and Tang to the renewed expansion of Islamic and Christian power after the 1500s. Hopkins identifies the "proto-globalization" period with the political and economic development that became especially prominent between about 1600 and 1800 in Europe, Asia, and parts of Africa. The third period, "globalization," is the colonial period from 1760 onward. Globalization that can be related to modernity started from 1800; it refers to the rise of the nation-state and the spread of industrialization. The last process, postcolonial or contemporary globalization, refers to the historical process from the 1950s till now. Hopper's (2006: 4) definition of contemporary globalization is the process by which the world has become interconnected, and that leads to the formation of global networks, transnationalism, deterritorialization, time-space compression, and the speeding up of everyday life. Held and McGrew (2007: 2–3) arrive at similar characterizations: "Globalization denotes the intensification of worldwide social relations and interactions such that distant events acquire very localized impacts and vice versa." Scholte (2005) further reports eleven categories or manifestations of contemporary globalization: communication, travel, productions, market, finance, organizations, military, ecology, health, law, and consciousness (for more details, see Scholte 2005: 49–84).

5.2.2 Globalization and Mediatized Society

Globalization involves the reduction of barriers physically, legally, linguistically, culturally, and psychologically and that makes people engage with one another without space-time barriers (Scholte 2005: 59). Local culture has been infiltrated by a new cognitive dimension such as knowledge and power, especially the soft power from major cultural product[3] exporters such as the US, Western Europe, Japan, and

---

[3]They are printed media enterprises, radio and television companies, news and features agencies, advertising and public relations firms, syndicates and independent companies producing and distributing print, visual and recorded material for print and broadcasting conglomerates, public or private information offices, data banks, software production, manufactures of technological equipment and so on. Productions from the communication industry are also known as the cultural industry because they record and reproduce cornucopia of social interactions, representations and organization systems in diverse media forms such as books, arts, films, recordings, television, radio, the internet, plays, concerts and music (MacBride 1980: 96–97).

the BRICS countries—Brazil, Russia, India, China, and South Africa— spurred by advanced communication technology. Soft power means these countries can influence the global public opinion and cultural values by using the media, not by military might (Shorenstein Center 2015). Sets of cultural products shared among many localities are called popular culture. It includes

> … the human activities, languages, and artefacts that grow and nourish people in communities and that generate observable, describable interest about its events and artefacts, within a community and between communities. (Holmberg 1998: 15).

New media technologies in this contemporary globalization period, such as Websites, Webpages, online chats, online videos, online TV, games online, and social networking sites such as Facebook, YouTube, Instagram, Tumblr, and Line, manifest the age of mediated society. The mass media and new media have helped create a virtual community and virtual experiences, at the same time they have been a woven thread of the globalization process, therefore, created a global consciousness, by which people compare their own living conditions with that of others (McQuail 2005: 51–52). Digital social communication, by using new information and communication technologies, has already blurred the boundary between the traditional "interpersonal communication" and "mass communication." We are now part of a digital network. We are mass media when we receive a message and send out to many people in a network. We also function as interpersonal communicators both offline and online, in actual or real time and perceived time. Through the mass and new media, people could see the otherness in the virtual reality and appropriate it to their local situation, according to Thompson (1995: 174). Moreover, a long-term process of the media influence on social change and institutional interactions has become noticeable. New media use in many countries has raised concerns about threats to national values, privacy, security, laws, and influences from major soft power producers in the world. The process whereby culture and society are increasingly dependent on the media and their logic, in such a way that the degree of the social interactions within a given culture and society modulated by media capital, is called mediatization (Hjarvard 2013:17). Mediatization is seen as a longer term process than mediation as the media's influence on the change of the social and institutional interactions is to be expected, whereas the mediation describes only the concrete act of communication by using a type of media in a given social context (Hjarvard 2013: 18–19).

5.2.3 Mediatization and Culture

Mediatization can have impacts on culture by increasing general information about the citizens of the world and cultures; accelerating contact with people who possess "the otherness"; enhancing contact with people within the same culture for support; redefining one's own identity and think of identity management; and widening the gap of access to communication technology, or the so-called digital divide (Martin and Nakayama 2004: 6). These aspects together with the changing of immigration patterns and demographics result in a culturally diverse society across the globe. This leads to the discussion on culture.

Culture is related to both material and non-material aspects. The material aspect concerns artifacts, heroes, and tangible products such as CDs, DVDs, and films. The non-material aspect is the socially constructed and subjective characteristics of culture. Brown (1995: 8–9) explains different elements of culture at three levels, starting from the shallow, the middle, and the deepest level, which interact among one another. Manifestation of culture are artifacts; language; behavior patterns in the form of rites, rituals, ceremonies, and celebrations; norms of behavior; heroes; symbols and symbolic action; beliefs, values, and attitudes; ethical codes; basic assumptions; and history. The shallowest level of culture is artifacts (stories, myths, jokes, metaphors, rites, rituals, and ceremonies), heroes, and symbols. The middle level of culture is beliefs, values, and attitudes. The deepest one is basic assumptions that concern the environment, reality, human nature, human activity, and human relationship. All of these levels are useful for cultural study within the tourism realm.

Another framework of culture, useful for analyzing culture, cultural diversities, cultural differences and similarities, is the one that Servaes (1999: 12) defines. He outlines four interrelated analytical components: a worldview (Weltanschauung), a value system, a system of symbolic representations, and a social-cultural organizational system. These four elements of culture are interrelated (Servaes 1999: 12). The first two elements are the immaterial parts of culture that can be manifested by the last two elements.

One other definition of culture which is useful for this study is the one defined by Agyeman (2013):

> predicated on differences and on otherness, and is a complex, dynamic, and embodied set of realities in which people (re)create identities, meanings, and values. Overlaying this is the reality of hybrid or multiple cultural and group affiliations. In this sense, no one person can be reduced to one single or fixed cultural form of identity.

This leads to the concept of cultural relativism. Cultural relativism is an assumption that there is no superiority among cultures and we should not compare cultures in terms of values or merits (Hall 1959). This assumption is not universally accepted as many communities still stick to the idea that their ways of life are better than others, which is called ethnocentrism (Mohammed 2011: 5). It is the tendency to judge "the other" in terms of our own experiences and expectations (Mohammed 2011: 5). Stereotyping is a way that one makes up a mental picture of groups and their supposed characteristics and one evaluates individuals from each group according to that mental picture, which can be a distorted view (Gannon 2008: 36).

Tourism and travel across borders, as influenced by the interconnectedness of this contemporary globalization period, bring people from other cultures to experience physically the otherness of another culture in a real-time situation. Tourist experience can be a valuable tool for cultural education, but at the same time it can come at heavy costs to the host culture and environment, known as cultural erosion (Mohammed 2011: 93). Culture relativity, ethnocentrism, and stereotyping as part of a cultural imaginary network on online discussion about Thai–Chinese culture will be analyzed under the framework of sustainable development.

5.2.4   Development Discourse, Sustainability, and EcoTourism

As the globalization process progresses, it is becoming clearer and clearer that inequalities exist in both degree and kind. Modernization or economic growth comes with globalization. However, globalization is not westernization as some parts of the world are not westernized but are still modernized, such as Singapore, Taiwan, and Iran (Nisbett 2003: 224). Modernization means transferring Western ideas and technological advances to Third World countries. Therefore, a new global system for transnational investments was established by a new transnational aristocracy that make countries in Latin America, Asia, and Africa become large importers of transnational goods (De Rivero 2001: 49–50). Not all parts of the world are equally modernized. In his book, "The Myth of Development," De Rivero (2001) discusses development and underdevelopment. De Rivero states that contemporary globalization is in fact

> the result not so much of free global competition among nations, but of a network of agreement and productive and financial activities among the transnational corporations (De Rivero 2001: 47).

He does not use the word "developing" because, in his view, this is just a eulogy of underdevelopment. Underdeveloped countries lack national capitalism; they have huge rates of unemployment and demographic growth; they export raw materials at unprofitable prices; and they have no choice but to seek transnational investments (De Rivero 2001: 47). Tourism is a way to earn incomes without much of investment. However, unplanned tourism can induce ecological disasters.

Ecotourism emerged under the framework of a new development paradigm, "Multiplicity," which emphasizes participatory-based advocacy communication for sustainability, sees development as a nonlinear process which can be planned to meet one country's need (Servaes 2013). Self-reliance, the ergonomics of using natural resources, autonomy, appropriate technology, participatory-based, and emphasis on sustainability are parts of the signature of the multiplicity paradigm (Servaes 1999). The need for development for sustainability is coined with the Thai perspective on development advocated by Payutto (1998), a renowned intellectual Buddhist monk. Payutto (1998: 183–184) emphasizes human development to be in line with natural environment conservation, not exploiting other humans or the environment, maintaining self-contentment, the wisdom to accept differences among mankind and understand that there are different paths to development with the aim of conserving the environment. Sustainable development in a Thai Buddhist perspective advocated by Payutto emphasizes three major components: economy, ecology, and evolvability (Payutto 1998: 168–173). Evolvability means the ability of humans to develop themselves to live in harmony with nature, not to conquer nature, nor destroy it. In his view, human development should come first on the development agenda.

Ecotourism is defined as "responsible travel to natural areas that conserves the environment and improves the well-being of local people" (TIES 1990). Principles of ecotourism meet the demand from a Thai Buddhist perspective on development. They are first, minimizing impact on the culture and environment; second, building environmental and cultural awareness and respect; third, providing experiences for both visitors and hosts; fourth, providing direct financial benefits for conservation of culture and environment; fifth, providing financial benefits and empowerment for the locals; and sixth, raising sensitivity to host countries' political, environmental, and social climate (TIES 2014).

These theoretical observations will be used as framework of analysis in this study.

5.2.5 Research Questions

From what has been described, three broad research questions can be drawn as follows:

1. Does Chinese tourism in Thailand meet the ecotourism standard and why?
2. Is there a way to a mutual respect for both the Thai host and the Chinese visitors? and
3. How to improve the tourism industry toward sustainability?

5.2.6 Methodology

A qualitative methodology is used to collect in-depth information. Samples of online media are purposely selected for this study. They are as follows:

1. "We Love Chinese Tourists" facebook page, which parodies news about Chinese tourists in Thailand.
2. Two episodes of Tang Kon Tang Kid (Each one thinks differently) program on YouTube discussing Chinese tourist problems.
   First episode, "Dramas from Chinese tourists" https://www.youtube.come/watch?v=qvPXIbWGg8Y, accessed July 15, 2015, discussed by President of Thai Travel Business, Mr. Sittiwat Shivarattanapon and President of Chinese speaking Tour Guides of Thailand, Mr. Marin Ruangwongsa.

Second episode, "Do Chinese tourists have no culture?" discussed by Ajahn Wiroj Tangwanich, an expert in Chinese Studies and President of the Thai Tourist Guides Association, Mr. Wiroj Litprasertnan. https://www.youtube.com/watch?v=g50OirEAUfDQ, accessed July 15, 2015.

3. Three episodes of "Kon Dang Nang Clear" (A Celeb comes to explain) program on YouTube discussing cross-cultural relations with China with a Thai expert in Chinese studies, Ajahn Wiroj Tangwanich. https://www.youtube.com/watch?v=50Wre7xiflU, accessed July 15, 2015.

4. Kom Chad Luek's "Chinese tourists in Thailand" program online. http://wwwl.youtube.com/watch?v=V20Ts7rFDHA, accessed July 15, 2015, discussed by Ms.Gigi Louis Kua, manager of Jiaranai Entertainment Ltd.; Mr. Woroj Litprasertnan, President of Thai Tourist Guides Association and Mr. Suriya Charoenphol, a Chinese instructor, Center for Tourist Guide training, Naresuan University.

5. The Chinese tourists accused of bad behavior in Thailand/Channel 4 news. https://www.youtube.com/watch?v=R2fbj3hepoE, accessed July 15, 2015.

6. Online newspapers that report about the Thai tourism industry and Chinese tourists, such as the Bangkok Post online, the British Dailymail online, and the China Daily.

To counter balance, discussions online about Thai tourists abroad have been studied as well. A few blogs on pantip.com were selected. They are as follows:

1. Do Thai tourists abroad behave the same way as the Chinese tourists? http://www.pantip.com/topic/31593232, accessed July 16, 2015;

2. For Thai tourists in Japan http://www.pantip.com/topic/31730724 and http://www.pantip.com/topic/31730724/page2, accessed July 16, 2015;

3. Silence and listen http://www.dailynews.co.th/article/283310, accessed July 16, 2015;

4. Ten Thai behaviors that you should not do when you are abroad, http://travelling-ok.exteen.com/20130313/entry, accessed July 16, 2015.

5.2.7 General Observations

A general description about the Chinese tourists derived from the key informants in the new media and social media under study reads as follows:

During the past two decades, large groups of Chinese tourists from Mainland China have used Thailand as their tourist destination. Popular destinations are Chiang Mai, the capital of the North, the coastal areas along the Andaman Sea and the South China Sea. The Chinese have chosen Phuket, Pattaya, and Bangkok as their destination for decades. In recent years, even more Mainlanders flock to Chiang Mai, mainly because of the success of the comedy film, "Lost in Thailand." Chiang Mai and Chiang Rai are relatively small and cannot cope easily with the huge numbers of Chinese tourists who came to visit. That creates culture shocks, both for the hosts and the visitors.

Behaviors of Mainland Chinese are distinctively different from those of Chinese from Taiwan, Hong Kong, Malaysia, Indonesia, or from other parts of the world, and even from the Thai-Chinese who migrated to Thailand for centuries. The post-World War socioeconomic development in China has made the culture of Mainland Chinese different from other Chinese diasporas. The country has leap jumped its own GDP to become top ranked in the world. However, the "cosmopolitanism" of the people still needs to catch up with the rapid leaps of economic advance. That may take decades. A Thai key online informant, who is expert in Chinese studies, states that Communist China cuts the Confucianism and Taoism elements in the cultural upbringing during the past 60 years and that has made the ethical base of Mainlanders different from Chinese diasporas and especially from the Thais, whose ethical base is based on Thai Buddhism. In other words, worldview, social values, and moral values of Chinese diasporas are different from one context to the another. The key informant states that Mr. Wen Jiabao already knew that having a lot of money alone does not gain respect from other countries. The good old Chinese culture from before the Cultural Revolution needs to be revived, he argued, and that may take years.

A general observation of those who have been to China is that the struggle for survival was high during the closed country period. Competitive and rushing behaviors are common now in postmodern China. Chinese people now have spending power, but still need to catch up on universal manners and etiquettes which may take time to learn. Hosts need to be patient. On the other hand, the Thai way of life is rather slow and less competitive. (One reason probably being, the country is fertile and has an abundance of food.) This may cause a culture clash between the Thais who were taught to be humble and mild and the Chinese who had to fight and strike for daily survival.

One of the first positive aspects about Chinese tourists that Thai online key informants mention is that Mainlanders spend a lot of money abroad. In Thailand, according to Thai sources online, each high-end Chinese tour group spends no less than 10 million Baht. Many key informants admit that the amount that Chinese tourists bring in is much higher than tourists from other countries. Mainlanders dare spend on luxurious items and on elite lifestyles during their vacation. Another positive aspect is that many Chinese tourists appreciate Thai people for having good manners and always smiling. One Chinese key informant reports that the Mainlanders who heard about the inappropriate behavior of some Chinese tourists in Thailand criticized their own people. They consider that, as a result, China loses face in the international arena. They called for strict measures to deal with those who behave inappropriately abroad. Some even suggest blacklisting or detention for offenders.

The distinctive stereotypes of Chinese from Mainland China exhibited in Thailand can be categorized as follows: (a) undisciplined behaviors: jumping queue, being loud, and littering; (b) unhygienic behaviors: spitting in public spaces, sneezing without covering mouth in public spaces and on food, urinating in public spaces or sacred areas, and unacceptable usage of public toilets (i.e., not flushing, not defecating in the toilet bowl, putting soiled toilet paper on the floor, and squatting on toilet seats).

Generally three types of Mainlanders come to Thailand: (1) private individuals; (2) high-end tourist groups (consisting mainly of well-educated people); and (3) the so-called zero-dollar tourists, which form the majority. This third type comes as part of a free of charge package tour (traveling, accommodation, and meals are paid for by the tour organizer). However, the tour organizers try to recuperate their expenses from getting commissions and from leading/luring the tourists to purchase souvenirs and luxurious items far more expensive than the actual prices. This is a kind of fraud that Thais are in complot with; it is an agreement between Thai tour agencies and their Chinese counterparts. This is not only zero but even lower than zero as the Thai partners often offer kickbacks to the Chinese counterpart to recruit "customers." Chinese customers who agree to come to Thailand, free of charge or even with promotions such as getting an air-conditioning unit, are mostly people from the remote areas of China, most often not well-educated and not used to traveling abroad and modern city life. They have saved up enough to come to Thailand, because it is cheaper than going to other places in China. They will be "herded" to where the tour guide wants them to spend money. Time allowed to use bathrooms and eat is limited, and that may be one of the reasons that some people do not pay attention to cleanliness and orderliness.

From the online discussions, four main problems can be identified regarding Chinese tourists in Thailand:

1. Thais have to blame themselves for not being able to screen the quality of Chinese tourists the same way, for instance, Bhutan does for it visitors. All key informants state that they really blame it to the management of this zero-dollar tourist issue.
2. The way Thailand has (mis)managed its tourism industry in general, and especially the Chinese component, over the past twenty years is part of the problem. Chinese companies and Chinese tour guides are allowed to operate freely without much Thai oversight or policy regulation. Many Chinese tour guides do not understand the Thai language and culture enough to explain to the tourists their faux pas or cultural differences between Thai and Chinese. There are not enough Thai guides who speak Chinese fluently. They cannot cope with the influx of Chinese tourists each year. The uncontrolled numbers of Chinese tourists (especially those without a tour guide) have damaged natural habitats, historical sites, and facilities. Some people write on walls and stones in forbidden areas, drive against the traffic rules, and exhibit unhygienic and disrespectful behaviors, such as urinating in sacred or public spaces. Thailand also still has a problem of not being able to manage the unregistered tourist guides. These are unqualified and untrained tourist guides who cannot advise tourists on Thai culture and travel.
3. As a consequence of what's stated under second, Chinese tour guides and shops do actually not bring in much revenue to Thailand. The money comes and goes

abroad (as the Chinese tour operators handle most of the cash flow). However, the hidden social costs and costs of conserving the natural resources and public facilities are being paid and maintained by Thailand.

4. Thai law is poorly enforced. Thai culture is easy going. Chinese tourists can get away with inappropriate and even illegal behavior. Thai may only being able to "complain" about it without much follow-up. However, it is known that the Chinese authorities have started penalizing Chinese tourists who committed unlawful behaviors. Even the tour guide of one particular group was suspended. The administrator of the Facebook page, "We love Chinese tourists," reports that certain behaviors which affect the public safety, such as driving dangerously or not abiding to traffic rules, and the public toilet misuse behaviors, compelled him to set up a page reporting these behaviors hoping that there would be a quick solution to the problems.

5.2.7.1 Cultural Relativity in Thai–Chinese Relations

Stereotyping is not a way to solve problems. It rather creates more hatred and discrimination. Cultural relativity is quite difficult as there is no universal standard to assess each characteristic. Here are a few examples discussed in our sample.

– On being loud: Chinese language sounds terse and abrupt. This is the linguistic nature. According to a Thai key informant who speaks Chinese and a Chinese key informant who speaks Thai, uttering sounds in Chinese is naturally louder than uttering sounds in Thai for a bilingual person. Thai tourists abroad behave the same way as Chinese tourists in this aspect, even though Thai language is softer and more musical in nature. Traveling in group gives power more than traveling alone and that makes some people realize less that they are disturbing others. According to data collected, big groups of Thai tourists are as loud as Chinese when traveling abroad.
– On etiquettes and cultural codes: Differences in culture do not mean that Mainlanders have no culture at all. Jumping queue is what Thais see Chinese tourists do. Queuing is not a Thai habit either. However, in the contemporary globalization period, some Thais are more exposed to Western cultural codes. Therefore, more and more Thais queue and wait in line.
  However, according to the blog traveling-ok, Thais were being reported to jumping queues abroad as well, especially in Japan. A blog advised Thais not to do the following 10 behaviors abroad: arriving late; jumping queue; sharing food/not finishing a dish/taking too much food from a buffet table/bringing food from outside to eat in restaurants; jaywalking; speeding/not abiding traffic rules/parking not in a designated area; not paying tips in a country where this is customary, like in the US; not wiping nose with a tissue; not cleaning up after using public facilities; no bargaining in a fixed-price place. This tells us that what is considered acceptable in one culture may not be acceptable in another

culture. Therefore, while traveling, one must know the culture of the host country.

– Authoritarian culture: Both Thai and Chinese share authoritarian characteristics in their culture and social system. Warnings or giving advice must come from elders. Chinese have lived under the polity of the Communist Party for many years. That makes them accept authority, such as police, more. Ekachai (July 5 2015) states boldly that Thai culture is deeply rooted in militarism, racism, hierarchy, and patriarchy.

– Saving face and losing face: This is a common characteristic of Asians. Both Thais and Chinese share this identity.

– Instant gratification: This is the consequence of mediatization. A new kind of individualism is the result of digital media usage, such as a short attention span, high-risk taking, experimenting, and self-expression (Elliot and Lemert 2006). The symbolic representation of this new individualism is manifested in the taking of "selfies" and the popularity of Instagram, Facebook and Line in Thailand, and their equivalents in China. In 2014, the volume of social sharing in China went up by 65% and the top ten social networking sites in China are, for instance, Sina Weibo (a hybrid of Twitter and Facebook Chinese microblogging Website), Renren (a Chinese version of Facebook), and Kaixin 001 (a cloned Facebook application for the Chinese market) (Sallee 2015).

5.2.8 Answering the Research Questions

5.2.8.1 Does Chinese Tourism in Thailand Meet the Ecotourism Standard, and Why?

By all means, it does not meet the ecotourism standard. Thinking along the modernization paradigm and "making a quick buck" is shortsighted. As there is more demand than supply; meaning that there are more tourists than the sightseeing sites can accommodate. The crowded tourists in a limited area, such as the temple of the Emerald Buddha or the Royal Palace in Bangkok, can damage the delicate antiques, decorations, or ancient walls. One key informant explained an incident when invaluable pieces of art were broken by Chinese tourists. Moreover, for the past 20 years, the management of the host country has not been efficient in screening the quality of tourists (i.e., the zero-dollar tours) nor unqualified and unregistered guides from China. Uneducated tourists do not help conserve the national heritage nor the environment of the host country. Littering and graffiti scribing are unlawful. Warning signs in Chinese and leaflets have been distributed for quite a while, but to not much avail. When natural resources or historical sites are damaged, there would not be more tourists coming. This is an example of cultural erosion.

5.2.8.2   Is There a Way to a Mutual Respect for Both the Thai Host and the Chinese Visitors?

Tour guides are important. They must tell the visitors what to do and what not to do before the visitors arrive at a sightseeing place.

Thais are taught to exclaim Poo-ke-I (phonetics for Do not do this!) when they see Chinese doing things that are not according to Thai customs. The administrator of the Facebook page, "we love Chinese tourists," reports that having such an online page gives access for Thais to complain about some Chinese tourist misbehaviors. However, that may lead to stereotyping of all Chinese which causes discrimination and hatred. He agrees with other online key informants that the Chinese tourists who create issues in Thailand are only a minority.

Thais and Chinese people are brothers and sisters. Many Thais are Chinese descendants. Mutual respect should begin with being polite and respectful to one another. Studying the host country's customs before traveling is necessary. Stereotyping of Chinese should be discouraged for Thais. The social media and edutainment media could help boosting mutual understanding between the two cultures.

5.2.8.3   How Can Thailand Improve on the Tourism Industry Toward Sustainability or Ecotourism?

Training more qualified Thai guides who can speak Chinese should be more emphasized. At present, only 2000 Chinese speaking Thai tourist guides "graduate" each year. Not enough to cope with the 5 million tourists a year. Thailand could offer to also train Chinese guides from China to understand Thai culture and customs better. The ethical conduct of tourist guides must be emphasized, and tight quality control of tourist guides should be implemented.

Thailand needs to stop the zero-dollar tourism scheme and collaborate more with four partners: Chinese tour agencies, Thai tour agencies, Thai government, and Chinese government. Screening for quality tourists is necessary. The number of tourists should be limited as well. Money cannot buy natural habitats and historical sites back. The Modernization paradigm that emphasizes growth and GDP from tourism as easy money should not be used as a framework. The multiplicity paradigm, which advocates respect for the local culture, preserving the environment and ecosystem as well as considering the social cost, should be taken into account instead.

Online discussants admit that Thais are not capable to manage visiting places, shops, or restaurants adequately. Handing out a queue card or arranging that people need to stay in line is not a big issue. Overseas companies and overseas guides that are not qualified must be banned. This would help the locals benefit from tourism. To organize ecotourism needs collaboration from many partners: the Tourism Authority of Thailand, tour agencies, schools and centers for training qualified tourist guides, government officials and employees of touristic sites, the local authorities, the entrepreneurs in the area, and the locals in the area of tourism. Participatory communication in a democratic fashion would work best.

As a short-term solution, handbooks of etiquette are being handed out to Chinese tourists to understand the Thai culture. Qualified tourist guides should be able to explain these. Chinese signs could help. Key informants state that if warning signs do not work, the hosts must be able to advise or call the police to advise the tourists. Chinese tourists respect the authority and warnings from people in authority better. Chinese discussants state that the online social network users in China would like to see Thai authorities to penalize Chinese tourists who behave illegally or inappropriately. By doing so, people will behave better next time.

For a long-term solution, both the Chinese authority as well the host country Thailand need to reach a mutual cultural understanding. Both Thais and Chinese need not to be judgmental. Illegal activities should be disciplined firmly and consistently.

5.3   Conclusion

Ad hoc problem solving is not the way forward toward ecotourism. Thailand should opt for ecotourism as to preserve and conserve the natural habitats for generations to come. Education is the key to it. If the locals were educated to take pride in their habitats, tourists would benefit and learn from the locals a great deal. Online discussions on social networking sites are healthy. It is a way of participatory democracy to learn and listen to opinions of the local as well as from Chinese representatives. A lesson to learn for both Thais and Chinese: money can buy comfort but it cannot buy "class." Hence, a paradigm shift from Modernization to multiplicity is needed.

With the advance of digital communication, clips and messages can be sent instantly. The media users should stop and think whether they are manipulated by hyperreality or not. Stop, listen, and think carefully while encountering any cultural similarity or difference.. Respond in a considered and respectful manner, and do not over-react. This will help strengthen Thai–Chinese relations.

References

Agyeman, J. (2013). *Introducing just sustainabilities. Policy, planning, and practice.* London-New York: Zed Books.

Brown, A. (1995). *Organisational culture.* London: Pitman Publishing.

De Rivero, O. (2001). *The myth of development.* London and New York: Zed Books.

Elliot, A., & Lemert, C. (2006). *The New Individualism: The Emotional Costs of Globalization.* London and New York: Routledge.

Ekachai, S. (2015, July 5). *Democracy? Don't hold you breath.* Bangkok Post online newspaper. http://www.bangkokpost.com/opinion/opinion/622868/. Accessed July 16, 2015.

Gannon, M. J. (2008). *Paradoxes of culture and globalization*. Los Angeles-London-New Delhi-Singapore: Sage Publications.

Grossman, N. (2015). *Thailand's sustainable development sourcebook*. Bangkok: Editions Didier Millet (EDM)

Hall, E. T. (1959). *The silent language*. New York: Doubleday.

Hjarvard, S. (2013). *The mediatization of culture and society*. London and New York: Routledge.

Held, D., & McGrew, A. (2007). *Globalization/anti-globalization*. Cambridge-Malden MA: Polity Press.

Holmberg, C. B. (1998). *Sexualities and popular culture*. Thousand Oaks: Sage Publications.

Hopkins, A. G. (2002). Introduction: Globalization—an agenda for historians. In A. G. Hopkins (Ed.), *Globalization in world history* (pp. 1–10). London: Pimlico.

Hopper, P. (2006). *Living with globalization oxford*. New York: Berg.

Lee, K. (2003). *Globalization and health. An introduction*. New York: Palgrave Macmillan.

Martin, J. N., & Nakayama, T. K. (2004). *Intercultural communication in contexts*. Boston: McGraw Hill.

MacBride, S. (Ed.). (1980). *Many voices, one world. Communication and society. today and tomorrow*. Paris: Unesco.

McQuail, D. (2005). *McQuail's mass communication theory*. London; thousand Oaks, Calif.: Sage Publications.

Mohammed, S. N. (2011). *Communication and the globalization of culture*. Lanham-Boulder-New York-Toronto-Plymouth, UK: Lexington Books.

Nisbett, R. E. (2003). *The geography of thought: How Asians and westerners think differently and why*. New York-London-Toronto-Sydney-Singapore: The Free Press.

Payutto, P. A. (1998). *Sustainable development* (3rd ed.). Bangkok: Buddhatham Foundation.

Sallee, H.(2015). *Social media in China. An Infographic*. Socialbrandwatch.com. http://socialbrandwatch.com/social-media-in-china-an-infographic/. Retrieved July 15, 2015.

Scholte, J. A. (2005). *Globalization: A critical introduction*. Basingstoke: Palgrave Macmillan.

Servaes, J. (1999). *Communication for development: One world, multiple cultures* (1st ed.). Cresskill, New Jersey: Hampton Press Inc.

Servaes, J. (2013). Communication for sustainable social change is possible but not inevitable. In: Sustainability, participation and culture in communication. Bristol: Intellect.

Thompson, J. (1995). *The media and modernity. A social theory of the media*. Cambridge: Polity Press.

Websites

Bangkok Post newspaper online. a. http://www.bangkokpost.com/news/general/469276/chinese-briefly-banned-at-white-temple. Accessed on July 3, 2015. b. http://www.bangkokpost.com/learning/learning-from-news/397926/chinese-tourists-invade-cmu. Accessed July 3, 2015.

China daily newspaper online. http://www.chinadaily.com.cn/china/2015-04/10/content_20400674.htm. Accessed July 3, 2015.

Dailymail.co.uk. a.http://www.dailymail.co.uk/travel/travel_news/article-2874902/AirAsia-flight-attendant-scalded-hot-water-noodles-unruly-passenger-wanted-sit-husband.html. Accessed on July 3, 2015. b.http://www.dailymail.co.uk/travel/travel_news/article-2966522/Presumed-Chinese-tourist-filmed-kicking-bell-Buddhist-temple-Thailand.html. Accessed on July 3, 2015. Accessed on July 3, 2015.

Shianghaiist.com. http://shanghaiist.com/2015/06/10/chinese_girl_publicly_urinates_in_t.php. Accessed on July 3, 2015.

Shorenstein Center on Media, Politics, and Public Policy. http://shorensteincenter.org/nachman-shai/. Accessed July 15, 2015.

The International Ecotourism Society (TIES). https://www.ecotourism.org/book/ecotourism-definition. Accessed July 14, 2015.

Tourism Authority of Thailand (TAT), 1995. http://expo.nectec.or.th/tat/stable/history.html. Accessed July 14, 2015.

Xinhua. http://news.xinhuanet.com/english2010/china/2011-06/28/c_13952856.htm. Accessed July 3, 2015.

Chapter 6
Self-Reliance and Sustainability from a Thai Perspective

Abstract This chapter discusses self-reliance and sustainability related to the Multiplicity Paradigm. The highlight of this chapter is the author's experience while attending two courses at Punpun organic farm in Phrao District, Chiang Mai, Thailand. The first course is a self-reliant course during which the author learned how to be self-reliant and self-sustained. The second course focused on organic vegan/vegetarian food making and appreciation. How all of these are related to the sustainable development and the multiplicity paradigm will be discussed in detail.

6.1 Introduction

Sustainable development has become a hot topic in Thailand. Networking of the locals who run organic paddy fields, establishing green hubs and learning centers, and producing herbal and organic products popular among Thai consumers are part of the sufficient economy scheme, introduced by King Bhumibol of Thailand. His Majesty introduced this paradigm in 1997 (Mongsawad 2010: 127). However, in those days, not many people paid attention to his suggestions. After Thais suffered from economic downturns, environment degradation, climate change due to global warming, and negative consequences of agro-industrialization, such as genetically modified products, contaminated foods, problems with seed conservation, and deforestation, Thais have started to appreciate a slower and healthier life. Anti-globalization is an ongoing process in Thailand. Therefore, this author, as a local Thai, is very proud to present the local perspective on development, sustainability, and self-reliance.

Let us start with the idea why we have to counter the stream of globalization. Can globalization be good for Thais at all? That leads us to our first topic.

6.2 Globalization and Sustainable Development

Globalization highlights the interconnectedness across boundaries, especially the monetary flows through free trades and loans from developed to developing countries. Unfortunately, the flow and networks that globalization characterizes do

© Springer Nature Singapore Pte Ltd. 2017
P. Malikhao, *Culture and Communication in Thailand*, Communication, Culture and Change in Asia 3, DOI 10.1007/978-981-10-4125-9_6

not bring along social justice, quality of life, fair global trades, equality, and environmental security to developing countries. Antidotes for social and environmental disasters within developing countries, including Thailand, could be promoting independence on foreign investment and foreign financial markets, redistributing land, democratic economic decision making, and the support of civil society organizations (Blewitt 2015: 24). All of these remedies are quite abstract and hard to implement as the dominant development paradigm used till now has been the modernization paradigm. This paradigm aims at Westernization, capitalization, and economic growth. Although globalization is not equal to Westernization, the main aim of this worldwide connectedness has been economic globalization. That brings more negative consequences than positive ones.

Blewitt (2015: 28–29) reports that positive aspects of globalization in the worldview of the market liberals are that it fosters economic growth, and with the applied science and technology it would improve the environment and our well-being in the future. In the view of these institutionalists, globalization speeds up chances for cooperation, capacity building, and innovation that is good for the environment. However, bad aspects that globalization brings in the eyes of these market liberals are receding economy brought on by market failures and poor governance that leads to market distortions. In the view of the institutionalists, weak institutions and poor global cooperation cannot help correct environmental disasters or counteract the ecological impacts caused by a nation-state. For bio-environmentalists, the bad aspects of globalization are the depletion of natural resources and unsustainable and ecological imbalances due to exploding economic growth, exponential population, unlimited consumption, and hedonistic materialism. According to the social greens, globalization has brought bad consequences on exploitation, social injustice and inequality, erosion of local-community autonomy, increase of drug-related global crime, human trafficking, and new forms of slavery, such as sweat shop workers, child labor, and prostitution.

All these bad consequences of contemporary globalization are indicators of unsustainable development grossly overlooked by the modernization paradigm. We have lost non-renewable energy and huge amounts of natural resources, and at the same time, we are faced with industrial residues which are toxic, and the use of even more chemicals to fight against insects and pests that have developed themselves to resist our conventional chemicals. Toxic gasses, such as carbon dioxide as a result of the combustion of fuels, fluorocarbon emissions from industries, deadly toxic gas from chemical warfares, and pesticides, insecticides, and herbicides used in agro-industries, cause the greenhouse effect, air pollution, and acid rains. The accumulation of the so-called greenhouse gases, such as carbon dioxide, methane, nitrous oxide, water vapor, and other gases, has caused the depletion of the ozone layer that shields the ultraviolet ray around the earth (livescience.com 2016). The ozone hole above the Antarctica and the melting of the icebergs on the North Pole are symbolic representations of global warming. Global warming causes seasonal changes and rapidly revives more growth of old extinction germs diseases such as malaria.

The shifting of the earth planes to accommodate the amount of water that increases from the melting of the icebergs results in tornadoes, cyclones, earthquakes, and tsunamis. Toxin released from industries causes air, water, and soil pollution that hastens the extinction of certain species. Desertification and deforestation out of greed and ignorance are also the cause of more natural disasters, such as mudslides and draught. Bacteria are more and more resistant to antibiotics.

Cross human and animal viruses appeared for the first time in this era. Though genetically modified organisms or GMO products were developed to cope with environmental problems, they may in the long run cause more disasters to human beings who consume this kind of food.

Sustainable development, in the view of Beckerman (2003: 7), has no clear conceptual basis. In many scholars' views, sustainability concerns many aspects: human rights, appropriate basic needs, a sound environment, considering the future of the offspring, and income distribution. There is, as of yet, no good enough measurable index according to economist Beckerman. This is the result of being trapped in the modernization paradigm that postulates that every variable should be quantified and measured. According to Beckerman, who cherishes economic growth the most, energy shortages and the depletion of non-renewable resources, extinction of certain kinds of animals, and dangers from climate change should not be taken too seriously because he believes that there will be a kind of compensation to balance out the loss in economic terms (2003: 9–41). This makes me think that the most unsustainable element of development is *our mind.*

According to many Western scholars, sustainable development has four main components: economic, environmental, cultural, and human. Without a good proportion of these four elements, we will not be able to reach sustainability. Current discourse on sustainable development involves sustainable yield, environmental sustainability, sustainable society, and sustainable development. To elaborate, Baker (2006: 26) explains:

Sustainable yield means maintaining the regenerative capacity of natural systems—for example, forests;

environmental sustainability: preservation of natural environment systems and processes, or addressing environmental issue to maintain social institutions and processes;

sustainable society: living within boundaries established by ecological limits, but linked with ideas of social equity and justice;

sustainable development: maintaining a positive process of social change.

From this view, we can see that sustainable development relies heavily on humans who can design social change in a desirable direction; who could safeguard the ecological balance; who could help prevent social justice from exploitation from other human beings. Humans need to be trained to have the right attitude toward the natural environment and conserve one's own natural resources. That is what we call "being educated." P.A. Payutto, a renowned Thai Buddhist monk, proposes that sustainable development in a Thai Buddhist perspective is comprised of three components: economy, ecology, and evolvability. In his view, the middle way or the balance of demands and resources has to be integrated (Payutto 1998: 168–173). His new term, "evolvability," means the ability of a human to develop oneself to live in harmony with nature, not to conquer the nature, nor to destroy it. In his view, *human development* has to come first. Educated citizens of the world need to have, first, the right occupation, meaning not to exploit other humans nor to the environment; second, a moral conduct to maintain self-contentment; and, third, the right wisdom to accept differences among mankind and understand that there are different paths to development with the aim to conserve the environment (Payutto 1998: 183–184). This view of Payutto strongly advocates the cultural perspective of development, that is, putting people at the center of development should be on the top of our development agenda. According to Payutto, greed (lobha), hatred (dosa), and ditthi (a domain comprising the following subsets: worldview, ideology, beliefs, religious beliefs, and social values) are three root causes of the environmental crisis and shortage of resources. According to him (Payutto 2007: 55–63), the beliefs or ditthi that hold control of world civilization can be categorized into three groups, as follows:

(a) The perception that mankind is separable from nature and can control or manipulate nature. Hence, the construction of dams, encroachment of forests, leaving carbon footprints, etc. are symbolic representations of this belief.
(b) The perception that fellow human beings are not "fellow human beings." Hence, slavery, modern day slavery such as child labor, migrant labor, forced prostitution, human trafficking, the demolition of the Native Americans or the Aborigines in Australia in the past, and racial discrimination in many parts of the world.
(c) The perception that happiness relies on material wealth possessions. This is one of the dimensions of development as growth already explained. This can create an issue of the have's discriminating against the have-nots.

It is not wrong to state that these ditthis that Payutto mentions are the morality behind the modernization paradigm.

Jan Servaes' (1999: 27–28) assessment of the modernization paradigm is worth reading. It can be summarized as mainly crunching the reality into numbers and drawing a straight line of cause and effect to variables which adopted the scientific methodology to study social science (also called positivism) hoping that it would work the same way everywhere. This view assumes that economic growth and urbanization, as in the Western world, will occur everywhere. It is the consequence of linear thinking and seeing the world as static. However, there are so many intertwined factors involved. The application of a laboratory experiment to the social world has resulted in environmental disasters in many developing countries. What happened with air and water pollution in China is a case in point. Aiming at raising per capita income without seeing the impacts, industrialization has brought to the exploitation of fellow human beings and non-renewable natural resources have proven to be short-sighted. Dumping industrial waste in water ways and releasing toxic gas into the atmosphere are what the so-called developed and developing worlds still do. Developed countries, under the disguise of globalization, turn the exploitation sites to developing countries. Thinking that what happens in one country will never have any impact to one's own has been proven to be not true. It is becoming abundantly clear that all man-made disasters affect people in the entire world. If the modernization paradigm, that has been the driving force for economic globalization for decades, has serious pitfalls, as stated above, what is our alternative?

Antidotes of the modernization paradigm are the dependency and multiplicity paradigms. While the dependency paradigm blames the external forces only about what went wrong in the developing countries, the multiplicity paradigm is an alternative framework that meets the demands for sustainable development.

6.3 The Multiplicity Paradigm

Keywords of this paradigm are interdependency, multidimensional development, respect of culture, satisfying basic needs, start from within a society or endogeny, self-reliance, ecological soundness, sustainability, participatory democracy, and structural and sustainable social change required. One can see that this paradigm covers all of the components needed for sustainable development. When we see that everything is interrelated, or in other words holistic, we can see many ways to achieve the same result. What constraints the way we choose can be a different context and culture of that society. We cannot just borrow the model that works in a Western country and use it without adapting it to our local context. This paradigm is close to the Buddhist way that sees that we have to be a part of nature and should live in harmony with nature.

Self-reliance in this paradigm, as defined by Servaes (1999: 79), is the ability to rely on strength and resources of a society with regard to its members' energies and its natural and cultural contexts. Sustainability, according to Servaes (1999: 79), is the interdependency of biosphere and ecosystems in a short, medium, and long term within local, national, or transnational levels.

6.4 King Bhumibol's Economic Sufficiency

The multiplicity paradigm is very much in line with King Bhumibol's Sufficient Economy Paradigm. King Bhumibol's paradigm is more practical and focused on Thailand, which is a tropical country. It is very suitable for the climate of Thailand as it has sunshine all year round. The land is fertile, and Thailand is in fact the bed of fresh foods for the world. The sufficient economy paradigm is therefore a new way of thinking and managing land by networking and cultivating human resources (a recent but rather a critical overview is available from Avery and Bergsteiner 2016).

King Bhumibol Adulyadej or King Rama IX, the current King of Thailand introduced this paradigm of sufficient economy to Thai people on December 4, 1997 (Mongsawad 2010: 127). He emphasized the middle way of thinking, living, consuming, and eating as a way to counter the globalization stream. This is in congruence with Buddhist teaching–the tisikkha: wisdom (panna), ethics (sila), and concentration (samadhi). True wisdom or *panna* is seeing through the reality of bias or prejudice. Ethics or *sila* is related to social justice and moral governance. Concentration or *samadhi* is about taking critical awareness of self to create peace, justice, and ecological balance (Sulak 2009: 45–46, 57–58, 94).

In a nutshell, it is about the middle path application of knowledge using knowledge, wisdom, and prudence; application of moral principles such as honesty, hard-working, sharing, and tolerance; and harmony, security, sustainability in people's lives, economic and social conditions, and the environment. All of this is within the context of globalization (Mongsawad 2010: 129). Realizing that globalization has its impacts, e.g., material, social, and environmental impacts, this sufficient economy scheme emphasizes self-reliance; appropriate technology; conservation of the natural environment; compassionate communities, i.e., helping one another in the agricultural production the traditional Thai way; production for own consumption then sell or barter the surplus; co-op building; capacity building and networking of the grassroots; and a new way of agriculture (Agri-nature foundation 2016).

To implement the sufficient economy paradigm, H.M. the King encourages Thai agriculturists to "grow what we eat and eat what we grow"; reduce the expenses as much as possible; and try to be self-reliant as much as one can. As summarized from the Website, Agri-nature foundation (2016), three steps of implementation are to be followed:

1. Produce agricultural products in a sufficient amount;
2. Join a group to produce, to market, to build up a strong community, and to engage in development;
3. Network with the public, the private sector, and the civic development sector.

Networking within the sufficient economy scheme consists of five major networks: Agri-nature foundation; organic farming of Thailand or Asoke network; BioThai foundation; the association of local wisdom representing the locals of the North-East; and the association of balanced agriculture-Taksom farm (Agri-nature foundation 2016). All of these networks have about 120 training centers around Thailand. The Website of Agri-nature foundation (2016) explains the nine-step training as follows:

1. Por Kin (having enough to eat—without using money). Thai agriculturists are encouraged to grow rice up to the amount that the entire family consumes yearly; grow fruit trees and vegetables so that they do not have to rely on the market economy. That implies also healthy eating. Thus, organic farming is recommended.

2. Por Chai (having enough to consume).
3. Por Yoo (having enough to live).
4. Por Rom Yen (having enough to feel cool).

   Steps 2–4 can happen together by growing forests. Growing forests will yield foods, clothing, and herbal medicine for people in a tropical country like Thailand. Besides, the forests yield wood to build homes and provide shade and shelter to the community. This will solve the problem of growing monocrops for money, which causes top soil erosion and drought as a result of no rains as a large number of trees were cut down to grow monocrops.

5. Bun. This is merit points one makes. By not possessing own assets but donating to temples and let the temples be centers of sharing.
6. Dana. This means learning to get rid of avarice and greed by starting giving and sharing within the community. Friendship and helping hands will be gained instead of monetary gains. In time of crisis, one needs this kind of relationship.
7. Keep. This means that farmers should store rice for their own year-round consumption, select, and keep 'quality seeds' for the next year. Also they are advised to keep preserved foods for the future.
8. Sell. Selling the surplus is what this guideline advises. Selling in this case is done with the feeling of "giving," meaning selling good quality products to the buyers.
9. Network to this revolution of the way people think and behave to solve the environmental crisis, epidemic crisis, economic crisis, and political/social crisis. The praxis of this paradigm has been implemented since 1988 on the land that H.M. the King requested that the Chaipattana Foundation purchased to be used as a demonstration field for integrated farming. The land of 32 rais (one rai is about 1600 m^2) was divided into two parts. The first part was to research on vegetables, herbs, fruit trees on the hilly areas, fragrant flower orchard, a fish pond, and a test of growing vetiver grass to prevent soil erosion. The second part was divided into the ratio of 30–30–30–10. The first 30% of the area was reserved for a pond to keep 18,000 m^3 water in the dry season. In the pond, the farmer can breed fresh water fish, such as Nile tilapias and silver barbs. The second 30% was stemmed for a paddy field which can be converted into a field of corn, bitter melons, and mung beans after the rice has been harvested. The third 30% of land will be used to grow fruit trees, such as mangos, guavas, and jackfruits; herbs such as kaffir lime and chili peppers; field crops such as sugar cane and bananas; and perennials such as acacias. The last 10% of land is to build a home, a stable for animals, driveways/walkways, and organic vegetable patches.

The sufficient economy scheme has been applied by a number of Thais, and it works. The case study I am going to report is my own experience at Punpun organic farm in Chiang Mai, Thailand, where I attended a self-reliant course from April 29 to May 2, 2016 and a vegetarian cooking course from May 6 to 8, 2016.

6.5   Self-Reliance and Sustainability from a Thai Perspective: Lessons Learned at Punpun Organic Farm, Phrao District, Chiang Mai, Thailand

6.5.1   Background of Punpun

The Punpun Center for Self-Reliance is one of the five initiatives listed as pioneers in the Thai organic community revolution in a recent sourcebook on Thailand's Sustainable Development (Grossman 2015: 259). Mr. Jon Jandai and his American wife, Peggy Reents, purchased about 10 acres of land in Mae Tang, Chiang Mai, and named it Punpun Organic Farm (Punpunthailand.org 2016a). After he had been toiling for seven years with hard work but without any savings in Bangkok, Jon decided to go back home in 1992 to Yasothorn Province in the North-East (Jon 2014: 9–10). He is a self-taught man who wanted to become self-reliant by working in his own organic paddy fields, cultivating organic vegetables, and breeding fish in a pond. Jon (2012: 28) claimed that he worked only 30 min a day but he could feed 6 members in the family. Moreover, he could sell the surplus of fresh produce and fish, and keep savings, which were good enough to purchase a piece of land in Mae Tang, Chiang Mai. Jon is well known in Thailand for his expertise in building adobe or mud homes, which he saw and studied by himself while traveling around in the so-called Four Corners: Utah, Colorado, New Mexico, and Arizona (Jon 2014: 16, 25). He started to be a volunteer instructor for mud-home building together with being an organic farmer in his own land. Later, there were more people coming to live with the family in the farm or purchase a piece of land adjacent to them and work with them. Up to now, about 15 permanent members make up the Punpun group (Punpunthailand.org 2016a, b). Jon stated in the introduction of the course that there are no rules and leaders. They live a simple life and share their labors, cooking turns, produce, and benefits. They established two restaurants selling organic vegan/vegetarian food and a cafe, using the produce from their own farm. Furthermore, they just launched a company called "Dhamma Thurakit" (Right Business) which will follow the sufficient economy scheme by being a middleman for a fair trade organic rice enterprise: buying rice from organic farmers, have it milled, and sell the rice to members. They bought a plot of land in San Pa Tong and will use that place to do organic paddy fields and train farmers to go organic as well.

They also sell products from their farm such as organic sun-dried bananas, peanut butter, organic kaffir lime shampoo, all-purpose liquid soap, home-made rice husk charcoal soap bars, and organic wild flower honey. The members share the profits from selling the products from the farm. Jon stated that they all have a salary of 6000 baht a month (@\$200) which they use to save for a vacation abroad. They also have their health fund for the members to use in the time of sickness so that everyone feels secured (Jon's lecture April 29, 2016). They use bicycles to move around most of the time.

They brainstorm and discuss ideas. They homeschool their children. Every member takes turn cooking three meals a day and babysitting. They all possess excellent cooking skills and creativity to turn everything produced by the farm into tasty and healthy food. They work from 8 am to 5 pm taking care of their farm and conserve non-GMO seeds and good local seeds. They give away free seeds to everyone who writes to them or comes to visit their farm.

They started to use their farm as a center of self-reliance and seed collection, as they are against GMO and the monopoly policy of seeds dictated by transnational and national agro-corporations. Later in 2015, when there was a drought and their farm ran out of water supply, Jon and Ek, his business partner, rented another plot of land of about 60 acres in Phrao district. They use this place to be a center of learning too. I was fortunate to be accepted to learn from them for the first self-reliant course ever launched at this center together with about 26 people, and about 20 people for the vegetarian cooking course.

The areas of teaching and learning are simple. The dining and food preparation area is a hall without walls—made out of wooden stilts and a tiled roof—with a cement floor and long wooden low tables so that students can sit on the floor. The kitchen is screened with a simple bamboo screen in the back. The kitchen consists of a charcoal stove and two gas stoves and many pots and pans and utensils. The area for yoga is just a roof on top of stilts among shady trees. A two-story building is a separate one with eight shower rooms and eight toilets using the water pumped from the water filtering units. The resident areas are separated for men and women. It is made of bamboo. It is just a roof on top of stilts with two separate platforms for sleeping in a mosquito net.

6.6 What I Learned from Punpun

During the self-reliant course, Mr. Jon was our main instructor. The course consists of introductory sessions on Punpun lifestyle and their philosophy; a tour of the farm; learning how to seed and sow, collect and preserve seeds; learning how to make compost; learning theory and practice of building a mud house, a water filtering system, making organic detergent, and EM liquid fertilizer; food and drink from the farm: bread making, cream cheese, vinegar, kombucha, and yoghurt; and discussions on health and basic yoga. For the vegetarian cooking course, other staff members were instructors. The course consists of making organic soy milk drinks, home-made tofu and food from soybean residues from the tofu making process, and other international foods made out of the produce of the farm and mushrooms, such as salads, home-made pastas, and basic tomato sauce.

What I learned from my participatory observation at Punpun:

1. Development

A development indicator in the opinion of Punpun members is whether that path makes it easy for all the society members to fulfill the four basic needs—food, home, clothes, and medicine—in order to be self-reliant or not. If not, that way of development is a failure. Self-reliance is closely related to a right kind of development. We should be able to rely on ourselves for these four basic needs. All basic needs should be accessible and cheap and simple so that we can call it development for self-reliance, which leads to a sustainable life (Jon 2014: 75).

Everyone has a right to own a home. A mud home is strong and keeps one cool in summer. Everyone can build it from the soil in their land, husk, clay, and water. They can use straw or even recycled materials, such as used tires and bottles. Jon states in his interview with Wirapha (2012: 28–29) that being able to build one's own mud home brings back power and pride to the underprivileged in a society. Development the way that is suitable for Thais is to have one rai (1600 m^2) and to grow food and live as explained in King Bhumibol's sufficiency scheme. Thais know how to weave cloth from cotton and silk. Thai traditional and alternative medicines are booming. Thai massage therapy and the development of Thai herbal medicine have been acknowledged and supported by the Thai authority.

What I learned here is to think out of the box. There is no standard and no right formula in nature. Just roll up your sleeves and observe nature to understand what element makes each plant grows. Do what we really like, and there is a way out of every dead end. To be able to make a choice makes life worth living. This leads us to the next topic.

2. Self-Reliance

According to Jon, self-reliance is freedom from being a slave of capitalism. Jon considers money not secured as it fluctuates and inflates. Food is the only security we have, states Jon boldly. He does not deny the importance of having some money to buy necessities but he thinks that living outside the system is best. He stresses that the traditional Thai lifestyle was a relaxed, warm, and sustainable one, but it was replaced by the crave for money and luxury money can buy. That causes stress, suffering, chaos, etc. The ability to stand on one's own feet is the ability to produce healthy food for consumption and knowing the local wisdom about herbal medicines and how to live a healthy life. Self-reliance does not mean one need to be a lone Don Quixote. Rather, it means you can be interdependent with others. It is a survival as well as an individual and as a group, Jon argues (2013: 105). There must be a way to live together and work together, to use brain power as well as labor work in a balanced way. That implies that self-reliance also means both group and individual autonomy and the ability to organize social supports. It also means living a simple life and being able to design our own life. The first step to self-reliance is lower down one's own consumption (Jon 2013: 107).

From my observation, self-reliance works here in any definition. Strategically speaking, this place has its own hot spring. They drilled wells and pumped (powered by solar energy) the underground water to fill up the moats they dug around the land. The underground water is being pumped through a filtering system that they invented and tested to get filtered water good for drinking. They also build toilets and bathrooms and use septic tanks, and use the residue to help make compost. They have 40 cows roaming the field. Ducks and chickens have their pond. They also dug a large pond for breeding fish. They grow bananas, fruit trees, vegetables, and rice without using chemicals. They make their own compost from straw, cow manure, grass and green residues from the garden, and kitchen leftover. They eat what they grow and they grow what they want to eat. They also buy or barter organic products such as organic soybeans with other organic farmers in their network. They are very good at adapting all they can find in the farm to make nutritious and delicious vegan/vegetarian food. They eat sometimes fish and eggs from their own farm. They all wear simple clothes and a pair of flip-flops. They practice yoga. The children receive love and care from the members. Everyone looks healthy and happy.

When people have freedom to design their own lives, they will have self-confidence and less fear. A step further is sustainability that I am going to address next.

3. Sustainability

According to Jon (2013: 76–79, 124–125), the index of economy growth is an indicator of self-destruction. We are lured to consume and think that happiness depends on more and more consumption. In fact, overconsumption burdens us and that takes a toll on our health. Moreover, safeguarding valuables create suffering. Worse than that, to be able to consume more, one needs to work more, which means no leisure time and no bonding in the family. Many Thais realize that we have enough quantity of food but not much variety as food has been monopolized by a few transnational corporations. The work that they do is to destroy natural resources to grow monocrops with chemical fertilizers and we are left with polluted and infertile soils.

Jon, in his interview with Songglod (2012: 79), indicates that Thailand had been sustainable before we changed to consumerism the Western way about 50 years ago. The old life is good; it just lacks the health dimension so we need to improve on it. When we discarded our old system and jumped right onto the bandwagon of modernization, we became slaves of the system right out and had no time to enjoy life.

Sustainability means a holistic way to live in harmony with nature, and that way of life will nurture our sustenance. Thinking that we are an inseparable part from nature, we will treat the nature with understanding and respect. As a consequence, we will have fertile soil and clean water to use. We will be able to consume healthy food from our own production. We can build our own home. We can be a part of a fair trade. We should be free from being a slave to money, a hierarchical system, amnaj (power), fame, etc. Sustainability also means that understanding that life is a cyclic order; we always have happiness that comes together with suffering. Happiness is not the most important thing in life. The understanding of oneself in life is more important, states Jon in his interview with Songglod (2012: 80). Jon gives an interview about his view on understanding oneself and life as follows:

Being sick is a phase of life that our body warns us to come back to ourselves.....

If we don't have a cold, we won't think about the value of being able to breathe.

If we are not sick, we won't know the value of being alive.

... Death may be the most precious gift the nature gives to you. But we are afraid of it. Because no one explains to us how it is like to be dead. If we accept this fact of life, our problems in life will diminish. The way I understand life is to reduce the fear of death.

Jon Jandai's interviewed by Songglod (2012: 85–86)

In our relatively short lifetime, we should appreciate the miracle of life that gives us biodiversity. Life's appreciation is more important than hedonism. More importantly, sustainability means freedom from being dominated by biogenetic engineering. That leads to my next topic.

4. The Importance of Seed Collection and Conservation

Jon has campaigned for a seed center. He states in his interviews and in the course that the disappearance of seeds is alarming. The attempt of a few international biotech industries to patent native seeds and living resources has made the locals rely on the seeds and chemical fertilizers they need to buy from these companies. Biopiracy and genetic engineering to control the seeds implies that the locals cannot collect seeds from their vegetables and fruits any further because they are either mutated or no longer re-plantable. That is what is being practiced in the contemporary globalization period. Genetically modified organisms to implant strange genes from other species to make living sources resistant to pests, insects, and regrowth are being promoted by multinational corporations. This kind of development reduces the biodiversity, and worse than that the safety of eating them. GMO products are not tasty. They are made to look good and grow fast. Food is developed for the market, not for the consumers.

There will not be self-reliance without the collection of authentic seeds that our ancestors have developed and selected for us. It would be a shame that we cannot transfer our indigenous wisdom of food selection, food safety, cooking, preservation of seeds, and more to the coming generation (Punpunthailand.org 2016b).

It is a very clever move that Punpun did not advocate to go against the multinational corporations overtly, but Punpun opts to give away seeds free of charge to anyone. This is to counteract the idea that everything is for sale. The way to conserve seeds is simple for Punpun people. They do not have a big cooling place to keep the seeds. They know that the seeds can be kept for a year in a refrigerator. They encourage people to sow the seeds, grow them, eat the produce, and if it is good, just keep the seeds and share them. By so doing, we keep the seed conservation in our lifestyle; quietly, we are no longer slaves of biotech industries.

5. The Importance of Networking with Like-Minded Others

Punpun has collaborated with other green and organic centers, such as the Santi Asoke, the agri-nature foundation, Mor Keaw (self-reliance in health care) group, Sufficient agriculture club, and more. The Santi Asoke group is an austere Buddhist group that prefers to call its profits from selling homegrown organic produce and products, such as herbal medicines, as bun-niyom (merit preferable), meaning that they do not count just on monetary gain as profits. They sell cheaper than the market prices for all necessities in their own shops spreading in major cities in Thailand. The agri-nature foundation trains the general public to the practice of sufficient economy model in line with the King's initiative. Herbal medicine, integration farming, organic farming, and more are the topics of their training. Mor Keaw is a leader in health care. He gives advice on Thai traditional medicines and coordinates centers of volunteers on alternative healthcare providers, seed collection, and databases for local biogenetics, for instance. The sufficient agriculture club gives advice to people on how to be successful organic farmers, provides seeds free of charge, and networks with people who do sufficient agriculture in 77 provinces. By collaboration and networking, Thais will make exemplary endeavors to be one of the leaders in organic and healthy food that could sustain the entire world.

6.7   Conclusion

At the end of this chapter, I cannot help thinking of a good friend of mine, Dr. Woraphat Phucharoen, about the title of his recent book, "Lighting the lantern is better than cursing the dark." This is what I would like to remind the reader…that living out of the system is possible. The dominant development paradigm can change if we change our worldview. Development for self-reliance and sustainability does not only apply to Thailand or other developing countries, developed countries can also learn from the developing. One does not need to be a Buddhist to have compassion and loving kindness to others. Compassion, loving kindness, rejoice with other's success, and equanimity can be learned and trained. Self-reflexivity is needed to see our own greed, lust, anger, want, and needs. Stop and think before you consume. Know that everything is interrelated. What goes around comes around so easily in a globalized world fueled by the Internet and social media. Monetary benefits made on the suffering of others are not sustainable happiness. Jon Jandai, a man who dropped out from a university because he thought it does not teach him how to live a sustainable life, is now a teacher of me who has a Ph.D. I learned a lot from him and from other Punpun staff members. This is a proof that a well-educated or civilized person does not need to be a Western person, a university degree graduate, or an avid reader. It is about whether one possesses high morality or not.

I firmly believe that the multiplicity paradigm provides a theoretical perspective for inner wisdom-growth and healthy environment, and that King Bhumibol's sufficient economy scheme is doable. I sincerely hope that there will be more green hubs, centers of learning, more organic farms, more co-ops, and more and more young people interested in conserving seeds and Thai wisdoms. It is not too late to roll up your sleeves and do the right thing!

References

Agri-nature Foundation 2016 http://agrinature.or.th/node/169, accessed on April 8th, 2016.
Avery, G., & Bergsteiner, H. (Eds.). (2016). *Sufficiency thinking. Thailand's gift to an unsustainable world*. Sydney: Allen & Unwin.
Baker, S. (2006). *Sustainable development*. London and New York: Rutledge.
Beckerman, W. (2003). *A poverty of reason. Sustainable development and economic growth*. Oakland, CA: The Independent Institute.
Blewitt, J. (2015). *Understanding sustainable development* (2nd ed.). New York: Routledge.
Grossman, N. (Ed.). (2015). *Thailand's sustainable development sourcebook*. Bangkok: Editions Didier Millet (EDM).
Jon, J. (2013). *Klab Baan (Return to home)*. Mae Tang, Chiang Mai: Punpun.
Jon, J. (2014). *Yoo Kab Din. 16 Pi Karn Pun Din Pen Baan (Living with soil. 16 years of moulding soil to become homes)* (3rd ed.). Mae Tang, Chiang Mai: Punpun.
Mongsawad, P. (2010). The philosophy of the sufficiency economy: A Contribution to the theory of development. *Asia-Pacific Development Journal. 17*(1), 123.
Payutto, P. A. (1998). *Sustainable development* (3rd ed.). Bangkok: Buddhatham Foundation.
Payutto, P. A. (2007). *A buddhist solution for the twenty-first century* (17th ed.). Bangkok: Pimsuay Printing.
Servaes, J. (1999). *Communication for development: One world, multiple cultures*. Cresskill, NJ: Hampton Press Inc.
Songglod, B. (2012). Interview with Jon Jandai. In *Ngarn Keb Maled Pun Kue Ngarn Sudtai Nai Chiwit Phom. (Seed collection is the last task in my life)* (pp. 58–87). Mae Tang, Chiang Mai: Punpun.
Sulak S. (2009). *The Wisdom of Sustainability. Buddhist Economics for the 21st Century*. Chiang Mai, Thailand: Silkworm Books.
Wirapha, U. (2012). Jo Baan Din who turned to become seed collectors. In *Ngarn Keb Maled Pun Kue Ngarn Sudtai Nai Chiwit Phom. (Seed collection is the last task in my life)* ((pp. 26–36). Mae Tang, Chiang Mai: Punpun.

Websites

Livescience.com 2016. http://www.livescience.com/37743-greenhouse-effect.html. Accessed May 27, 2016.
Punpunthailand.org 2016. a. http://punpunthailand.org/index2f07.html?page_id=62. Accessed May 28, 2016. b. http://thai.punpunthailand.org/indexa123.html?p=631. Accessed May 29, 2016.

Chapter 7
Mindful Communication and Journalism from a Thai Buddhist Perspective

Abstract From a Buddhist perspective, social change needs transformation of both spiritual and personal aspect. Mindful communication is a methodology to a critical self-reflexivity and social engagement. It is to help us cope with greed, anger, and delusion. For the mass media and the media industry, delusion deems to be addressed for both media practitioners and corporations. Being mindful is being addressed in this chapter. Anyone, not only journalists, should benefit from mindfulness understanding and practice.

7.1 Introduction

Digital literacy is what being discussed as the utmost necessary characteristics of a postmodern individual. It captures the ability to effectively use computer applications, video and internet tools to communicate and use information for individual and social purposes. Positive use of digital media is discussed as important for lifelong learning, economic productivity, and democratic engagement (Wilhelm 2004: 117), and negative use of digital media as for increasing greed, anger, and delusion (Loy 2008: 92–93). Loy (2008) discusses in detail how media corporations maximize their profits by using advertisements to entice consumerism and distort the worldview of consumers to believe that personal image and worth are judged by the ability to consume, not mention about the ability to aware of social injustice and engage in social responsibility. Loy calls the role of the media as institutionalized delusion. The process of delusion can be called mediatization. It is a process whereby culture and society are increasingly dependent on the media and their logic in such a way that the degree of the social interactions within a given culture and society modulated by the media capital can be observed within social institutions, between institutions, or in a society (Hjarvard 2013: 17). Spurred by communication technology in the few last years of the twenty-first century as one of many factors, mediatization is seen as a part of globalization characterizes the postmodern culture industry. Print media, electronic media, and new media including social media are our cultural products that we create and share.

© Springer Nature Singapore Pte Ltd. 2017
P. Malikhao, *Culture and Communication in Thailand*, Communication, Culture and Change in Asia 3, DOI 10.1007/978-981-10-4125-9_7

With the breakthrough of the new media as a consequence of the digitization revolution, new formats of self-expressions have become popular as the audience gains recognition of his/her private, social, and public achievements (Hjarvard 2013: 150). Hjarvard (2013: 11) explains that the mediatization process affects an individual autonomy and social belonging in such a way that the individual gains more autonomy by relying deeply to the external world in the act of connecting to the available large social networks. He calls this phenomenon, *soft individualism*. Elliott and Lemert already observed a new kind of individualism in 2006. They propose that globalization has a profound impact on the individual level. They define this *new individualism* as a highly risk-taking, experimenting, and self-expressing individual underpinned by new forms of apprehension, anguish, and anxiety. High levels of individualism can lead to *narcissism*. Twenge and Campbell (2009: 19) state in their book, "The Narcissism Epidemic," that the central feature of narcissism is a very positive and inflated view of self and this value is growing rapidly in the American culture fueled by the mass media, including the new media, and changes in parental approaches to upbringing that emphasizes self-expression. Symbolic representations of the new American culture of self-expression or participating audience/amateur journalist are the emphasis on celebrities in the media, the success of Facebook as a social networking site, the uploading of personal videos on YouTube, twitter (micro blogging and text-based social networking or SMS on the internet via its own website), and blogging (Twenge and Campbell 2009). In many cases, the audience can be a target or a commodity when the abundant of profit-making orientation and values are built into the media system in advertising (Jackson et al. 2011: 63). The self or ego of the audience in these cases will be coupled with commercial products to increase self-confidence, self-respect, self-esteem, etc.

It is not wrong to conclude that mediatization and Western idea of practicing journalism are to feed the perpetual perception of self. We may be aware that someone else is suffering than us from reading news. We may feel sad about the tragedies in the news, but at the same time, we may feel relieve that we are not the victims. We look at entertainment and we feel happy we could forget our own boredom, pains, and suffering for a while. We escape from our self and back to our own life. Escapism is what the media help us for a short while. Then, we are back to the real-world again.

7.2 How Thai Buddhism Helps Us See Through the Perception of Self!

Mindful communication and journalism are based on the Buddhist phenomenology and philosophy. It helps global citizens and journalists alike as observers to communicate effectively with the observant while framing the observed.

Mindfulness is clear comprehension or being aware of our own thinking, verbal actions, and deeds. Practicing mindfulness helps us control our five sense doors: eyes, ears, nose, mouth, and skin in such a way that our mind will not fluctuate according to any sensory pleasure. Foundations of mindfulness or Satipatthana 4 are as follows: mindfulness of the body, mindfulness of the feelings, mindfulness of the mind, and mindfulness of dhammas (what Buddha taught).

Mindfulness of the body means practicing to just know every movement of the body and the in- and exhales. It is a practice of not thinking about self (I, me, and mine). Mindfulness of the feeling means practicing to just knowing the present feeling whether it is pleasant, painful, or neutral. Mindfulness of the mind means keeping up with the mind whether it is sad or clear or with defilements. Mindfulness of dhammas means contemplating on mind objects, knowing all the hindrances of a clear mind and the feeling of self (Gotoknow.org).

According to Kittiwat (2007), mindfulness of the body includes observing and reflexing on our breaths, postures, understanding, impurities (or the aging process) of the anatomical parts, and the four elements of the body (earth, water, air, and fire element). Mindfulness of feelings includes contemplating on feelings caused by external factors and within the body. Mindfulness of the mind includes contemplating on the mind with greed, anger, delusion, sloth, developed or undeveloped, and free mind from all defilements. Mindfulness of dhammas includes contemplation on five hindrances (from awakening), five aggregates, sex internal and external sense base, and seven enlightenment factors or Bojjhanga 7. Five hindrances are sense-desire, ill will, sloth and torpor, distraction, and worry. Five aggregates consist of corporeal body or rupa, feeling or vedana, perception or sanna, mental formation or sankhara, and consciousness or vinnana. Six internal and external sense bases are eyes and the visible objects; ears and sounds; nose and odors; tongue and tastes; the body and tactile objects; and the mind and mind objects.

Mindfulness is a very important base to reach enlightenment. The Bojjhanga principles that Buddha laid to reach the enlightened mind which is the mind that is free from "self-attachment" are what Bhikku P. A. Payutto (2013: 64–87) explains its seven steps as follows:

(1) Sati sumbojjamkam (mindfulness). It is having consciousness to fixate the mind on what we are dealing with or what we are facing.
(2) Dhammavijaya sumbojjamkam (investigation-of-states). It meaning use wisdom or investigative mind to peruse the true nature of things that mindfulness brings to us. Loy's (2006: 47–51) social roots of suffering should be considered. He refers to institutionalized greed of our economic system, the institutionalized ill will of the collective aggression in war fares, and the institutionalized delusion of the advertising, the stock market, and the ignorance of the crimes, violence, exploitation, poverty, and injustice in the world.

(3) Viriya sumbojjamkam (energy). This means the forthrightness to face obstacles, hard work, and dangers.
(4) Piti sumbojjamkam (rapture). It is the state of letting go of worries, anxiety, and annoyance.
(5) Passatthi sumbojjamkam (tranquillity). When the mind is at ease, the body will be calmer and relaxer.
(6) Samadhi sumbojjamkam (concentration). This means the concentration process that our mind fixates to what we are doing. We will have strong willpower to do things.
(7) Ubekka sumbojjamkam (equanimity). This is the neutralized mind, not being disturbed by positive or negative triggers or stimulators.
    (Pali terminology and its English translation is from http://webboard.watnapp.com/forum.php?mod=viewthread&tid=511. Accessed on July 23rd, 2016).

By being aware of our inhaling and exhaling breaths (mindfulness of breathing), we live in the present moment. Walking meditation, another practice of living in the present, is a synchronization of our body, feelings, and mind by concentrating on the movement of our feet. By so doing, we are cultivating awareness and concentration. Insight meditation (the technique Buddha taught called anapanasti), is another way of training the body, feelings, mind, and dhammas by gaining mindfulness and investigate the ti-lakkhana (three characteristics of existence): dukkha (suffering), anicca (impermanence), and anatta (not-self).

7.3 Not-Self

When Prince Siddhartha Gautama, who passed away 543 years before Christ, discovered how to attain Nirvana (supreme bliss), he proclaimed himself as Buddha, which means the "awakened" or the "enlightened" one. According to the Buddhist doctrine, the perception of "self" brings sufferings to us because we all are ignorant (avijja) in the true nature of things (May 1984: 95). That is the fact that there is nothing we can hold on to as we all age; go through phases of change, get sick; and eventually die either of old age, sickness or accidents. The world keeps changing and nothing is here to stay. Impermanence or transience (anicca) is evident. For most Buddhists, it is known that the Lord Buddha taught that there is no substantially self (anatta) and the doctrine of anatta had become a dogma and a component of Buddhist identity (May 1984: 93). By adhering to the thirst or craving for existence of self and sense experience (in Pali, tanha), suffering (dukkha) arises (May 1984: 9).

Suffering arises because of one's clinging to the five aggregates or Panca-khandha that give us the individual illusory of self (or atta): matter (rupa), feeling (vedana), perception (sanna), mental formation (sankhara), and conscious-ness (vinnana). Then, come the clinging and attachment to "self." The "self" or ego is being fed by tanha (craving): craving for sensuality (kama tanha), craving to be …(bhava tanha), and craving not to be …(vibhava tanha); and defilements (kilesa). The Lord Buddha explains that because of the feeling of "self," we cultivate 10 kinds of unwholesome quality of the minds or kilesa: greed(lobha), hate(dosa), delusion(moha), conceit (mana), speculative views (ditthi), restlessness(uddhacca), shamelessness (ahirika), and lack of moral dread or unconscientiousness (anot-tappa). To elaborate, because of the feeling of existence of self or delusion of self, one's perception that one is important gives rise to conceit. The perception that one's point of view is very important gives rise to a speculative view. Anything that is not pleasurable for self or for self's view point could create annoyance or anger. Not having what one wants creates lust or greed. The mind would go restless, and the self may commit any unlawful deed out of shamelessness and lack of moral dread.

According to the doctrine of causal genesis or dependent origination (Paticcasamuppada), the ultimate cause of ego or self can be traced through 12 conditioned factors that we fabricate our "self." The way to cease the "self" is that we ought to see that the world is full of complexity. There is no linear thinking or direct causality. Traditional Pali interpretation of Paticcasamuppada is as follows: With ignorance (avijja) as condition, fabrication or mental formation (sankhara) arises. With mental formation, consciousness (vinnana) arises. With consciousness as condition, mind and matter (namarupa) arises. With mind and matter as condi-tion, sense gates/contact via seeing, smelling, touching, tasting, and hearing (phassa) arises. With senses gates as condition, feeling (vedana) arises. With feeling as condition, craving (tanha) arises, and with craving as condition, clinging (upa-dana) arises. With clinging as condition, becoming (bhava) arises. With becoming as condition, birth (jati) arises. With birth as condition, aging and death (jar-amorana) arises.

This interpretation implies the cycle of past, present, and future life or reincar-nation. However, Buddhadasa rejects this view of Paticcasamuppada. The cessation of self can happen any moment of practice in our life time. From his arduous Pali studies and interpretation of many ancient books, he came up with a new inter-pretation as seen in the following chart.

7.4 Twelve Causally Linked Stages of the Paticcasamuppada

Diagram constructed by Patchanee Malikhao from source: (Jackson 2003: 111–119)

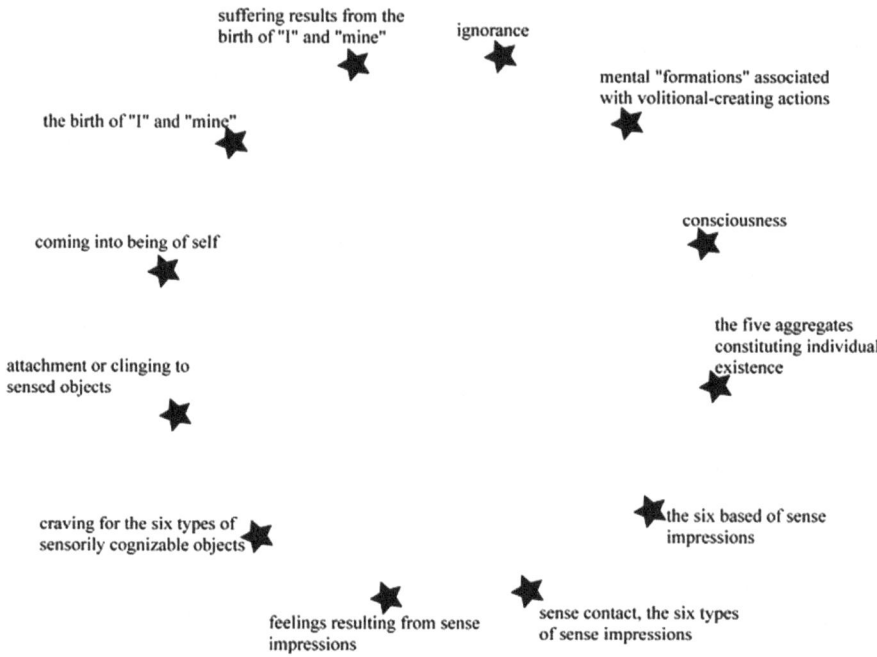

Buddhadasa Bhikkhu, the renowned Buddhist Thai monk, explained his interpretation of the cycle of Dependent origination as follows:

"It (Paticcasamuppada) is a detailed demonstration of how suffering arises and how suffering ceases. It also demonstrated that the arising and ceasing of suffering is a matter of natural interdependence. It is not necessary for angels or holy things, or anything else to help suffering arise or cease….

….The other aspect of Paticcasamuppada is that it demonstrates that there are no sentient beings, persons, selves, we or they here or floating around looking for a next life. Everything is just nature: arising, existing, and passing away" (Buddhadasa 2002: 23–24). This interpretation of Buddhadasa boldly contradicts the traditional interpretation, which was widely accepted that the birth of I and mine meant the birth of a human life, the suffering of the birth of I and mine meant the old age, sick, and death. Together with this law, the ways to the cessation of "self" was explained by Buddha as the *four Noble Truth*. Masao Abe stated the essential of what Buddha taught as follows:

The four Noble Truth, the fundamental teaching of Buddha, run as follows: that existence is suffering; that the cause of suffering is craving or thirst; that by the extinction of craving nirvana may be attained; that the means for the attainment of nirvana is the practice of the Eightfold Noble Path: right view, right intention, right speech, right conduct, right livelihood, right effort, right mindfulness, and right concentration. (Abe 1989: 205)

The first Noble Truth is Dukkha or suffering because of "self." The second Noble Truth is Samudaya (the origin of suffering) as explained in the 12 interdependent conditioned of causality or Paticca Samuppada. The third Noble Truth is Nirodha (the cessation of suffering and its causes) and the eight magga (8 paths to cease suffering).

Nirvana (Sanskrit spelling) or Nibbana (Pali spelling) is the existential awakening to egolessness, the non-attachment to the dualistic view of suffering and pleasure (Abe 1989: 205–296). Nirvana is a state of attaining sunyata (emptiness) of the mind after one has trained one's mind to overcome craving, which is the root cause of pleasure-suffering (Thomas 1951: 96). According to Macy (1984: 80–81), the very core of Buddha's original preaching is as follows:

> Craving (tanha, lit. thirst for existence) is pierced (patviddha, mastered) by him who knows (janato); for him who seeth (passato) naught remains (n'atthi kincanam). To see in the sense that matters for liberation is not to "grasp" or "understand" the infinite of truth; rather than difficult achievements are the natural corollary when naught remains of the impurities and lusts (kilesas) generated by craving. Enlightenment and liberation condition one another: "seeing" and "knowing" in this sense is equivalent to freedom from the bonds of sense impressions and resultant desires.

The concept of sunyata or voidness or naught appeared in the prajnaparamita or the perfection of wisdom as the antithesis of "self" (Thomas 1951: 96). (Loy 2008: 48) explains "shunyata" (his spelling of sunyata) as the absence of self-existence. By explaining sunyata, Loy infers to the interdependence of all beings and things and the theory of relativity. By realizing that things have no essence of its own, we should not attach to anything, not even the concept sunyata as Buddha compared sunyata to a raft that takes us across the ocean (of suffering from greed, ill will, and delusion) (Loy 2008: 49). That means we need to let go, even the raft that we ride on.

Buddhadasa famously stated that "Nirvana here and now" meaning that one can attain nirvana in this life time as one can be trained to be a selfless person. Working with the voidness of self is appreciated as one does not put own bias or predisposition in seeing the truth or in the intervention of social interactions.

7.5   Mindful Communication

The essence of mindful communication is the realization of suffering of self and the collective suffering of the commons. In so doing, we need to cultivate spiritual development, regardless of our religious beliefs. We all can be a kind-hearted person as The Dalai Lama (2006: 21) states:

> When we take into account a longer perspective, the fact that all wish to gain happiness and avoid suffering, and keep in mind our relative unimportance in relation to countless others, we can conclude that it is worthwhile to share our possessions with others. When you train in this sort of outlook, a true sense of compassion—a true sense of love and respect for others—becomes possible. Individual happiness ceases to be a conscious self-seeking effort; it becomes an automatic and far superior by-product of the whole process of loving and serving others.

Thus, mindful communication calls for nondiscriminatory and nonviolent social change. Adhering to the Precept Five (refrain from killing, lying, cheating on partner, stealing, and taking intoxicants) may not be enough to understand the complexity of the socials that contemporary globalization brings. McLeod (2006: 159–163) recommends 14 precepts of modern Buddhism or the fourteen mindfulnesses of the order of interbeing:

(1) "Aware of the suffering created by fanaticism and intolerance;
(2) Aware of the suffering created by attachment to views and wrong perceptions;
(3) Aware of the suffering brought about when we impose our views on others;
(4) Aware that looking deeply at the nature of suffering can help us develop compassion and find ways out of suffering;
(5) Aware that true happiness is rooted in peace, solidarity (groundedness), freedom, and compassion, and not in wealth or fame;
(6) Aware that anger blocks communication and creates suffering;
(7) Aware that life is available only in the present moment and that it is possible to live happily in the here and now;
(8) Aware that lack of communication always brings separation and suffering;
(9) Aware that words can create suffering or happiness;
(10) Aware that the essence and aim of a sangha is the practice of understanding and compassion;
(11) Aware that great violence and injustice have been done to our environment and society;
(12) Aware that much suffering is caused by war and conflict;
(13) Aware of the suffering caused by exploitation, social injustice, stealing, and oppression;
(14) Aware that sexual relations motivated by craving cannot dissipate the feeling of loneliness but will create more suffering, frustration, and isolation."

These are the modern Buddhist Precepts that we should vow not to continue them.

These are dhamma for awakening so that we use own wisdom to see through things with a calm mind without bias or prejudice. Thanissaro Bhikku (2004: 186) explains five principles that Buddha taught to honest acceptance of blame in order to communicate and reconcile with others in a community:

1. "We are always responsible for our conscious choices;
2. We should always put ourselves in the other person's place;
3. All being are worthy of respect;
4. We should regard those who point out our faults as if they were pointing out treasure; and
5. There are no—repeat, no—higher purposes that excuse breaking the basic precepts of ethical behavior."

By practicing the four principles of Brahmavihara: metta (loving-kindness), karuna (compassion), mutita (sympathic joy), and ubekka (equanimity), one should also be able to communicate with others fruitfully and successfully. This means one should let go of egoism, racism, ethnocentrism, institutionalism, regionalism, nationalism, etc. to see others as friends who have to face all sorts of sufferings as well as one does. One should cultivate the mind to have empathy. One should be able to congratulate others on their success. And one should have a steady and neutral mind, not being fluctuated by envy, ill will, anger, or delusion.

A practice of being selfless by walking or sitting meditation helps us get rid of our ego, so that we are aware that all of us are interrelated. Speaking and treating others with kindness and practice being selfless is important to socially engage with others.

7.6 Mindful Journalism

In order to distinguish professional journalists from citizen journalists in a globalized world, the professional journalists should empower and mobilize the socials as change agents. Not only must professional journalists be multi-skilled or well-rounded, but they must also be able to think critically. As journalism was invented in the Western world, Western ethics and epistemologies influence the codes of conduct and practices of journalists. Practicing journalism as an occupation of constructing the reality has encountered a number of discourses such as empiricism versus interpretivism; objectivity versus subjectivity; relativism versus idealism; and interventionism versus non-involvement (worldofjournalism.org).

As the world is mediatized, demands for revisiting the functions those journalists perform to mediate the reality, the media, and the audiences have been subjects for discussion. Those functions are the populist disseminator, the detached watchdog, the critical change agent, and the opportunist facilitator (Hanitzsch 2011: 477). New demands for new genres of journalism are more pronounced such as public journalists, media affiliated political commentators, and journalists on religion. As the audience is no longer considered a passive one, they have more chances to participate in the interactive Internet world. Traditional functions' journalists perform such as to inform, to be a watch dog, to entertain, to form the public opinion, and to educate may not be enough. It is expected that journalists in the postmodernity should serve the public; address fluxes of challenges posted by new economic, technological, cultural, and social realities; and be innovative, self-reflexive, and agile in social negotiation of information and knowledge (Servaes 2009: 371–374).

Knowing that dukkha is caused by clinging or attachment to the self or ego (uppadana), tanha (craving), kilesa (defilements), or avijja (ignorance) would help journalists to report news and features that cause less suffering to the audience. How applying anatta or no-self framework in reporting the news will be elaborated. How objectivity can be addressed from the doctrine of anatta and Paticca Samuppada will also be addressed.

7.6.1 Selfless Journalists as Actors/Agents/Propellers for Social Interactions and Social Change

Should journalists report truth or fact as it is or should they act a social actions interpreter depends on their epistemologies. The first one is called empiricism and the second one interpretive social. Journalists may take a distance and report what they see as a detached watchdog or populist disseminator by reporting about celebrities, any self-inflated occurrence such as gossiping on stars, and new fashions. But do these roles help the journalist to propel the public to think critically about sustainable development?

Marshall Singer clarified that we experience everything in the world not "as it is" but only as "the world comes to us through our sensory receptors" (Singer 1987: 9). It is not wrong to say that everyone's world is unique and different because of the influence of one's own culture and the meanings of symbols we interpret. Singer then concluded that perceptions of reality are more important than reality itself because we all only experience "learned external world" (Singer 1987: 35). Singer's view is in line with the interpretive social epistemology. Objectivity does not exist in this epistemology there is no truth out there to grasp. However, subjectivity or interpretation of reality does not mean that journalists have to relinquish the principles of protection of sources, accuracy checking, and writing and constructing of story (Kumar 2012: 57–59). Gibson (2004: 32) reports that day-to-day journalists should not make any claim to "objectivity" which requires presentation of all relevant facts, asking and answering relevant and significant questions, and comparing and testing competing views.

I would like to interpret Gibson's objectivity as impartiality.

As the notion of the media and journalist is not quite precise as in the past because the participating audience can perform the act of journalism as well (Ward 2011: 5), practicing "selfless journalism" is advisory to construct less biased interpretation of reality. By doing so, a selfless journalist should be able to conform to what Rajesh Kumar proposes about journalists as agents of a global public sphere as:

a well-informed, diverse, and tolerant global "infosphere" that provokes citizens to become engaged in issues and provides a counterbalance to the lies of tyrants and the manipulation of information by special interests (Kumar 2012: 33).

As the world is full of suffering, sensational journalism should be avoided to increase a higher degree of self-indulgence or individualism in the society. According to Kumar (2012: 33, 38), Journalists should help report non-slanting information and avoid fueling conflict or xenophobia as well as promote a global ethics, serve the citizens of the world, and promote non-parochial understanding.

Selfless journalists should avoid the following biases:

(1) commercial bias: means biased toward money-making business;
(2) temporal bias: biased toward the immediate (ever-changing cover story even when there is little news to cover);

(3) visual bias: biased toward visual depiction of news;

(4) bad news bias: biased toward selecting of only bad news;

(5) narrative bias: biased toward a full story of event that has to have beginning, middle, and end;

(6) status quo bias: this means news media never questions the structure of the political system;

(7) fairness bias: compelled to get reaction from an opposing camp;

(8) expediency bias: competing for freshness and timely of news and never comes to a rest; and

(9) glory bias: journalists asserting themselves into the story they cover (especially TV reporters) (Kumar 2012: 63–66).

The mind of the selfless journalist is the mind that bears the voidness of bias, impartiality, egoism, ethnocentrism, hatred, and delusion. It should be the mind of compassion (metta) and loving kindness (karuna), ready to sympathize or empathize others (mutita) and stay undeterred of lust and craving (ubekka or equanimity).

They should, out of compassion and loving kindness to the other beings who are subject to suffering, raise many critical issues such as the indifference as the consequence of being exposed to hyperreality, the decadence of hedonism, and the moral bankruptcy of consumerism. They should, as well, require the possession of mutita (sympathic joy) and ubekkha (equanimity) to understand and stay neutral while working and facing sufferings in the media context. First, there are internal issues of the media such as the copyright issues, social network issues (such as exposing self and risking privacy), violent entertainment, and pornography. Second, external issues are such as crime, poverty, immigration, racism, drug use, economic depression, and epidemics (May 1973: 201).

Moral conducts of selfless journalists can rely on the wholesome course of action (Kusala-kammapatha) as follows:

> bodily action: abstention from killing; abstention from taking what is not given; abstention from sexual misconduct verbal action: abstention from false speech; abstention from tale-bearing; abstention from harsh speech; abstention from vain talk or gossip mental action: non-convetousness; non-illwill; and right view (palikanon.com).

Bearing in mind that there is no right or wrong, no objective reality, and no way to link the individual to the universal (Gibson 2004: 47), a selfless journalist should be able to cultivate strong moral sensitivity and ethical conducts as well as sound knowledge of interdisciplinary fields.

7.7 Conclusion

Some people think that Buddhism is a religion of negativity. It denies sensual pleasure, sex, violence, and even the existence of self. In fact, being Buddhist means being awakened. Anyone can benefit from the Buddhist ways of meditation and philosophy without having to convert to Buddhism. Meditation is the strength that Buddhism offers world citizens to observe our own mind and feelings and learn to change attitude and behavior. Mindfulness can be practiced so that one is at peace with heart filled with less greed, ill will, and delusion.

Mindful communication is essential for global citizens to let go with those -ism, be it egoism, institutionalism, nationalism, etc. Kindheartedness can be cultivated. All citizen journalists or social media communicators should practice mindfulness so that there will not be defamation, false news, hoax, vulgarism, and invasion of other privacy online. Mindful journalism is essential for journalists to observe whether they feed delusion to the audience. By being mindful, we can cultivate peace, nonviolence, loving-kindness, compassion, and understanding of others. Sadhu Sadhu Sadhu.

References

Abe, M. (1989). *Zen and western thought* (W. R. LaFleur, ed.). Honolulu: University of Hawaii Press.

Buddhadasa, B. (2002). *Paticcasamuppada Thammasapa, Bangkok* (1st ed.). Bangkok: Thammasapa Publishing.

Elliot, A., & Lemert, C. (2006). *The new individualism: The emotional costs of globalization.* London and New York: Routledge.

Gibson, D. (2004). *Communication, power, and media.* New York: Nova Science Publishers Inc.

Hanitzsch, T. (2011). Populist disseminators, detached watchdogs, critical change agents and opportunist facilitators: Professional milieus, the journalistic field and autonomy in 18 countries. *The International Communication Gazette, 73*(6), 477–494.

Hjarvard, S. (2013). *The mediatization of culture and society.* London and New York: Routledge.

Jackson, P. (2003). *Buddhadasa, theravada buddhism and modernist reform in Thailand.* Chiang Mai: Silkworm Book.

Jackson, J. D., Nielsen, G. M., & Hsu, Y. (2011). *Mediated society.* Ontario: Oxford University Press.

Kittiwat, K. (2007). One only way. http://oneonlyway.blogspot.com/2007/04/blog-post_12.html. Accessed July 23, 2016.

Kumar, R. (2012). *The culture of journalism: Value, ethics and democracy.* Sumit Enterprises: New Delhi.

Loy, D. R. (2006). WEGO: The social roots of suffering. In M. McLeod (Ed.), *Mindful politics. A buddhist guide to making the world a better place* (pp. 45–54). Boston: Wisdom Publications.

Loy, D. R. (2008). *Money sex war karma. Notes for a buddhist revolution.* Boston: Wisdom Publications.

May, J. D. (1984). *Meaning, consensus and dialogue in buddhist-christian communication: A study in the construction of meaning.* Berne-Frankfort on the Main-Nancy, New York: Peter Lang Publishing Inc.

McLeod, M. (2006). The political precepts. The fourteen mindfulness of the order of interbeing. In M. McLeod (Ed.), *Mindful politics. A buddhist guide to making the world a better place* (pp. 159–163). Boston: Wisdom Publications.

Payutto, P. A. (2013). *Chue Karm-Roo Karm-Kae Kar (Believing in Karma-Knowing Karma-Resolvig Karma).* Nakorn Pathom, Thailand: Wat Yanavatesakawan

Singer, M. R. (1987). *Intercultural Communication: A Perceptual Approach*. Upper Saddle River, NJ: Prentice Hall College Div.

Servaes, J. (2009). We are all journalists now! *Journalism 10*(3), 371–374.

Thanissaro, Bhikku. (2004). Reconciliation, Right & Wrong. http://www.accesstoinsight.org/lib/authors/thanissaro/reconciliation.html. Accessed April 9, 2017.

The Dalai Lama. (2006). A new approach to global problems. In M. McLeod (Ed.), *Mindful politics. A buddhist guide to making the world a better place* (pp. 17–28). Boston: Wisdom Publications.

Thomas, E. J. (1951). *The history of buddhist thought*. London: Routledge & Kegan Paul.

Twenge, J. M. & Campbell, W. K. (2009). *Living in the Age of Entitlement. The Narcissism Epidemic*. New York: Free Press.

Ward, S. J. A. (2011). *Ethics and the media. An introduction*. Cambridge University Press: Cambridge.

Wilhelm, A. G. (2004). *Digital nation. Toward an inclusive information society*. Cambridge, MA: The MIT Press.

Website

Gotoknow.org. https://www.gotoknow.org/posts/401104. Accessed July 23, 2016.

Palikanon.com. http://www.palikanon.com/english/wtb/g_m/kamma_patha.htm. Accessed March 26, 2014.

Wordofjournalism.org. http://www.worldsofjournalism.org/pilot.htm. Accessed March 25, 2014.

Chapter 8
Human Trafficking in Thailand: A Culture of Corruption

Fiona Servaes

Abstract Human trafficking continues to be a significant problem in Thailand. Only recently, mainly due to international media coverage and policy decisions made by foreign governments, the Thai government has started to address the issues. Although small advances are slowly being made, the underlying root causes within Thai culture and governmental corruption continue to undermine anti-trafficking efforts, and can therefore only be solved if significant structural changes are being made.

8.1 Introduction

Since the 1980s, with the rise of the sex industry in Thailand due to the Vietnam War and the discovery of the AIDS epidemic concurrently, human trafficking has been on the global agenda of high politics. It has quickly become recognized as the third most profitable criminal activity in the world (Rafferty 2013), generating billions in illegal profits per year. In Thailand especially, human trafficking continues to be a significant problem, as Thailand is a major source, destination, and transit country for forced labor and sex trafficking. Although Thailand has sought to improve its trafficking image and reputation of the country's infamous sex industry over the past few years, sex trafficking cases and the numerous unprecedented reports and investigations of slavery within Thailand's seafood industry in recent years continue to dominate international attention (International Labor Rights Forum and the Migrant Workers Rights Network 2016). Outcries from activists and critics have solely been addressed recently by the Thai government due to the influx of international media coverage in the past two years. Although advances are slowly being made, the underlying root causes within Thai culture and governmental corruption continue to undermine anti-trafficking efforts and must therefore be addressed in order to make significant changes.

The original version of the chapter was revised: Missing author name has been updated. The erratum to the chapter is available at 10.1007/978-981-10-4125-9_10

© Springer Nature Singapore Pte Ltd. 2017
P. Malikhao, *Culture and Communication in Thailand*, Communication, Culture and Change in Asia 3, DOI 10.1007/978-981-10-4125-9_8

8.2 Sex Trafficking in Thailand

The rapid expansion of sex tourism in Thailand is largely to blame for the beginning of the sex trafficking issue. The Thai government poured millions of dollars in the country to promote tourism in the 1980s. Sex massage parlors had already become popular during the Japanese occupation during World War II, and, afterward, during the Vietnam War in the late fifties, sixties, and early seventies, US military servicemen used Thailand as a "rest and recreation (R&R)" destination. This generated a booming sex industry in the cities. By 1982, sex tourism had become Thailand's top foreign exchange earner (Jayagupta 2009).

Several cultural factors support the continuation of Thailand's commercial sex industry in Bangkok, Chiang Mai, Pattaya and other urban and tourist areas, which involves recruitment of mainly girls and women from the poorest regions of the north of Thailand and the Greater Mekong Subregion countries: war-torn Cambodia, Laos, China, Myanmar, and Vietnam. Deep cultural practices exist within family and kinship relations that encourage sex trafficking of women and girls in Thailand and Southeast Asia in general.

Shame is used as a weapon; it is culturally embedded and used to manipulate girls (Jeffrey 2002; Long 2004). Women's sexuality and bodies are treated as commodities, and traditional values place a high value on purity and virginity, as is evident in traditional practices such as bride payments and dowries, which allow an increase in the bride and her family's social and economic status. Mail-order brides are also common in Thailand, creating the opportunity for the brides to be exploited by the husbands (Long 2004).

Although prostitution has been illegal in Thailand since the 1960s, the law has consistently been ignored. Due partly to the enormous amount of money generated in the sex industry, prostitution has become "socially accepted" in Thai society, with prostitutes facing a lesser degree of stigmatization than in other countries. Traditionally, children of Thai parents have an obligation to help their family financially. Thus, if prostitution is the only way to ensure the family's well-being and status, prostitution is simply seen as a job. As human trafficking is a process that typically begins with recruitment by false promises of good jobs and higher wages, victims fall prey to the practice as they seek enhanced economic opportunities to better their quality of life. Many trafficking victims enter the sex industry as sex work provides material wealth, allowing poor and uneducated persons to generate more income than other wage labor available to them. According to the International Labor Organization (International Labour Organization 2014), adult sex workers are able to earn more than the average monthly salary in Thailand, often making them the main wage-earners in their families.

It is also quite common in Thailand for existing sex workers to act as recruiters and have a cascading deceptive recruitment process involving a gradual socialization into selling sex. An example of this is that it is common for the workplace to simulate family sociality and have a fictive family ethos of a family home. The venue owners often view the recruitment process as a form of helping sex workers out of poverty to rationalize their own conduct. Managers of venues catering for sex commerce thus typically use fictive kin-terms in everyday speech, with female managers commonly referred to as "Mother," and male managers as "Father." Additionally, the managers refer to the sex workers empathetically as their own children.

8.3   Men as Sex Trafficking Victims

As most of the human trafficking victims in the world are women and girls trafficked for the purpose of sexual exploitation, there is a highly gendered approach to the study of human trafficking, as can be seen in Thailand's situation. Most of the academic research on human trafficking has been focused on sex trafficking from feminist perspectives, with the study of prostitution being closely related (Piper 2005).

Academics argue that boys and men have so far solely been addressed in labor exploitation, rather than sexual exploitation (Glotfelty 2013). Although 20% of human trafficking victims in the world are boys and men, because of the gendered approach on human trafficking, little research has been done on the inclusion of men and boys in sex trafficking and this issue remains predominantly anecdotal (Piper 2005). There are gaps in understanding and acknowledgment of boys as trafficking victims, due to a degree of social tolerance. Mainstream concepts of human trafficking deem that females are more vulnerable to sex trafficking and thus are in greater need of legal protection than males. This traditional narrative obscures the plight of male victims, especially male children, who remain invisible victims (Jones 2010).

The absence of publicity surrounding boys also allows criminal networks that specialize in obtaining young boys for sex and pornography to be increasingly attractive and easy to target. Particularly with the popularity of social media and online technology, traffickers are no longer limited to luring, recruiting, and selling victims in the streets. Traffickers now have multiple quick choices to target thousands of people through Facebook, Instagram, Twitter, Snapchat, WhatsApp, etc. Traffickers are also able to control victims using remote surveillance (Whiting 2015).

As gender stereotypes are obscured by modern-day media perceptions of male dominance, showing immunity to victimization, males are believed to be strong and less vulnerable. Consequently, it is problematic when the males do experience exploitation because most men do not view the exploitation as an issue that should be spoken about. They remain silent about their experiences because of socially imposed pressures to be self-reliant and desire sex, which results in lack of proper treatment as well as lack of social awareness on this issue. Not to mention that the effects on boys' development can be dire, as experiences of sexual violence in childhood hinder physical, psychological, and social development. Children who are abused sexually are also at heightened risk of being re-victimized, "either again as children or later as adults". Yet there prevails to be an alarmingly large lack of research and media attention on boys as trafficking victims. In fact, data on forced sexual acts among boys are only available for four countries in the world so far (UNICEF 2014).

Thailand is of course not included as one of these four countries, despite Chiang Mai, a big city in the north of Thailand, being a particular hub for boy victims. Many people in the north of Thailand live in hill tribes and rural areas where they unfortunately do not have proper government identification to be able to gain legal employment. Boys are thus often trafficked to tourist centers like Chiang Mai, to work in four major sectors of the male sex industry: show bars, freelance bars, karaoke bars, and male massage parlors (Glotfelty 2013).

After seeing many of the vulnerable boys in Chiang Mai, and realizing that there were no services offered to them, Alezandra Russell started Urban Light, a non-profit organization located just a few blocks from Chiang Mai's red light district, dedicated to empowering the lives of boys who are the victims of human trafficking (urbanlight.org). At the Urban Light Youth Center, boys aged 14–24 are provided with multiple services such as emergency housing, transitional housing, meals, health services, education, vocational training, and alternative employment. The need to help boys who are sex trafficking victims is clear, but Urban Light is still the only organization in Thailand that recognizes this need (more details in Servaes 2015). The skewed gendered approach to researching human trafficking ignores the hundreds of thousands of boys and men who are stuck in this cruel industry due to social beliefs, lack of education, poverty, and corruption.

8.4   Child Trafficking in Thailand

Although child trafficking in Thailand is not as well known as in neighboring countries like Cambodia (Boden 2012), there is no doubt that child trafficking victims can be found in Thailand's brothels, bars, hotel rooms, massage parlors, karaoke lounges, and private residences. Approximately 53% of identified trafficking victims in Thailand are children (U.S. Department of State 2016), with many trafficking cases reportedly facilitated by family members and friends. As a result, trafficking cases frequently begin as voluntary migration, with those most vulnerable being foreign migrants, children from hill tribe communities and other ethnic minorities in northern Thailand (ECPAT International 2016). Children between the ages of 15–17 are most involved in the Thai sex industry (ECPAT International 2015), and many children are forced to work in domestic servitude in urban areas or forced by parents and brokers to sell flowers and beg on the streets. Those most vulnerable are children from hill tribe communities and other ethnic minorities in northern Thailand (ECPAT International 2016).

However, there are gaps in Thailand's monitoring of commercial sexual exploitation of children because of lack of law enforcements, largely due to corrupt police, prosecutors, and judges. For instance, police rely heavily on NGOs and INGOs to find and rescue sexually exploited children (ECPAT International 2016). As a result, there are also limited specialized services for child sex trafficking victims (Trafficking in Persons Report 2015) and the justice system is not child-sensitive. As explained in ECPAT International's Global Study on Sexual Exploitation of

Children in Travel and Tourism, it is common in Thailand "for child victims to be confined after their rescue in highly restrictive and inadequate shelters for long periods of time, often for the duration of the criminal investigation and prosecution."

8.5 Forced Labor in Thailand's Seafood Industry

The Thai human trafficking issue that has garnered the most international attention lately has been labor trafficking in the seafood industry. Migrant workers are most impacted in this industry, as 90% of the workforce in the seafood sector in Thailand comprises of migrant workers who often lack official legal identification and immigration documents. In 2014 and 2015 there was an inundation of media coverage on the egregious labor rights abuses in the largely unregulated fishing industries in Thailand (International Labor Rights Forum and the Migrant Workers Rights Network 2016). Tremendous numbers of Thai, Cambodian, and Indonesian men were reportedly trafficked on fishing boats off Thailand, treated as animals, and forced to work without pay and under threats of extreme violence. These men work in the production of seafood, mainly shrimp, for major retailers in the USA and Europe, such as Walmart, Carrefour, Costco, and Tesco. A six-month investigation by the *Guardian* found that enslaved men were living under "horrific conditions, including 20-hour shifts, regular beatings, torture, and execution-style killings. Some were at sea for years; some were regularly offered methamphetamines to keep them going. Some had seen fellow slaves murdered in front of them" (Hodal et al. 2014).

The Associated Press and other media sources followed a year later with more reports revealing that these inhumane conditions remain a significant concern, such as reports about slavery in Thai shrimp peeling facilities. Most notably in May 2015, reports flooded international news coverage about thousands of Rohingya Muslims who were uncovered in mass graves at trafficking camps on the Thai/Malaysia border (Equal Rights Trust 2014; Chambers 2015). Significant international response led to low rankings on the Trafficking in Persons (TIP) Report, a "yellow card" from the European Commission, lawsuits against Costco and Nestle seeking "forced labor" disclosure labels, calls for investigations into Thailand's shrimp industry and boycotts by US senators of Thai seafood products (International Labor Rights Forum and the Migrant Workers Rights Network 2016).

8.6 TIP Report

The annual Trafficking in Persons or TIP Report is the "world's most comprehensive resource of governmental anti-human trafficking efforts" which ranks all countries' governments on their effectiveness in tackling human trafficking issues based on a three-tier system. The Thai government continues to insist that they are working diligently to stop human trafficking, though little improvement has been made over the years. After being warned by the US Department of State that not enough was being done to combat human trafficking in the country, the 2015 TIP Report downgraded Thailand to Tier 3, stating that the Thai government "does not fully comply with the minimum standards for the elimination of trafficking, and is not making significant efforts to do so." By placing countries on Tier 3, its lowest possible ranking, countries can face many consequences under US law, including non-trade-related sanctions, restrictions on US foreign assistance and disqualifications to global financial institutions such as the World Bank.

Although the TIP Report is widely used by international organizations, foreign governments, and non-governmental organizations to compare anti-trafficking efforts of all countries and identify places where help is most needed, critics argue that it is merely a tool to publicly shame countries. After the 2015 TIP report was released, Bangkok protested publicly to Washington, requesting a change in ranking (Chuensuksawadi 2015).

Despite not demonstrating progress in combatting trafficking, Thailand was surprisingly upgraded to the level 2 watch list in the 2016 TIP Report, marking a rare boost for USThai relations (Bangkok Post 2016b). However, numerous investigations and reports, many led by The Guardian (Hodal and Kelly 2016), have found that all the trafficking and corruption issues of previous years still pervade Thailand's fishing industry. International rights groups argue that the US Government's choice to remove Thailand from the list of worst human trafficking offenders is merely politically motivated for its own strategic interests (EJF 2016).

In the report, reasons for improvement include claims that more cases are being investigated, more suspects are being investigated and corrupt officials will now face disciplinary action and life imprisonment. However, Phil Robertson, Deputy Director of the Asia Division of the Human Rights Watch said that "... leaders in Bangkok should recognize that the international community is not yet convinced of Thailand's commitment to fully implement anti-trafficking measures, or reduce the vulnerability of migrants to abuse that lies at the heart of Thailand's trafficking problem" (Bangkok Post 2016a).

8.7   How to End Corruption in Thailand?

Thailand has passed many anti-trafficking laws in the past, though weak institutional structures and corruption of government and law enforcement officials are rampant, including instances of police involved in the operations. Thailand ranks 102nd (among 175 countries) on the International Corruption Perception Index (2013). Grossman (2015: 154–157) calls it "a deep-rooted menace in Thai society." An astonishing 71% of the police, 68% of the political parties, and 58% of public officials/civil servants are seen as corrupt by the Thai public (Transparency International Survey 2013). Corrupt officials and policemen are known at times to act as kingpins, agents, and entrepreneurs in the sex services trade, working with traffickers and even engaging in commercial sex acts with children (Trafficking in Persons Report 2015). In other words, efforts to eradicate trafficking and slavery are detrimentally impeded by government complicity in trafficking crimes.

Thai law on child pornography also provides little protection to child victims, as there is no definition of child pornography and no prohibition and punishment for those who disseminate offer or possess child pornography (ECPAT International 2011).

Furthermore, the government prosecutes journalists and advocates for exposing traffickers, thereby attempting to hide aspects of the human trafficking issue.

Government corruption and lack of enforcement of anti-trafficking laws are not just issues in Thailand. By analyzing 76 variables on data from all countries of the world, Bales found that governmental corruption is a key predictor in determining trafficking *from* a country. In determining the factors that drive trafficking *to* a country, Bales found that the strongest predictor is the destination country's male population in the 60+ age bracket, followed by the level of governmental corruption (Bales 2007). This analysis suggests that "reducing corruption should be the first and most effective way to reduce trafficking," as corruption in both politicians and law enforcement officers only contributes to the lack of accuracy in information on human trafficking and the ease of transportation and exploitation of victims.

Transparency International (2015), in a special report on ASEAN, observes that each ASEAN member state has taken some steps to addressing corruption at the national level, such as ratifying the United Nations Convention against Corruption (UNCAC), but much more is urgently needed to stop corruption. If economic integration is not built on a strong foundation of transparency, accountability, and integrity, then the ASEAN community's vision will be jeopardized. A robust strategic regional anti-corruption framework through the formation of an ASEAN Integrity Community is therefore a critical step.

References

Bales, K. (2007). What Predicts Human Trafficking?. *International Journal of Comparative and Applied Criminal Justice*, (online) *31*(2), 269–279. Available at: http://www.tandfonline.com/doi/abs/10.1080/01924036.2007.9678771. Accessed June 6, 2016.

Bangkok Post. (2016a). US promotes Thailand out of Tier 3. *Bangkok Post*, June 29, 2016 Available at: http://www.bangkokpost.com/news/politics/1022357/us-promotes-thailand-out-of-tier-3. Accessed August 4, 2016.

Bangkok Post. (2016b). Thailand gets upgraded in US TIP report. *Bangkok Post*, July 1, Available at: http://www.bangkokpost.com/news/security/1024141/thailand-gets-upgraded-in-us-tip-report. Accessed August 4, 2016.

Boden, A.L. (2012). Human trafficking in Cambodia. *Report from the Princeton University Office of Religious Life*. Available at: https://lisd.princeton.edu/sites/lisd/files/Boden.pdf. Accessed August 4, 2016.

Chambers, P. (2015). Thailand must end its own Rohingya Atrocity. Bangkok needs to address its brutal mistreatment of the persecuted minority. *The Diplomat* Available at: http://thediplomat.com/2015/10/thailand-must-end-its-own-rohingya-atrocity/. Accessed August 4, 2016.

Chuensuksawatdi, P. (2015). Govt trafficking responses familiar. *Bangkok Post*, June 18, 2015 Available at: http://www.bangkokpost.com/archive/govt-trafficking-responses-familiar/595992. Accessed August 4, 2016.

ECPAT International. (2015). *Situational analysis of the commercial sexual exploitation of children in Thailand*. (online) Bangkok. Available at: http://www.ecpat.org/resources. Accessed July 14, 2016.

ECPAT International. (2016). *Offenders on the move: Global study on sexual exploitation of children in travel and tourism*. (online) ECPAT International. Available at: http://www. ecpatusa.org/global-study/. Accessed July 6, 2016.

EJF (2016). EJF calls for Thailand to remain on Tier 3 in 2016 US Department of State Trafficking in Persons Report. *Environmental Justice Foundation*. June 20, 2016. Available at: http:// ejfoundation.org/EJF-calls-for-Thailand-to-remain-on-Tier-3. Accessed August 4, 2016.

Equal Rights Trust. (2014). *Equal only in name*. Bangkok: The Institute of Human Rights and Peace Studies, Mahidol University, The Human Rights of Stateless Rohingya in Thailand.

Glotfelty, E. (2013). Boys, Too: The Forgotten Stories of Human Trafficking. https://www. fairobserver.com/region/asia_pacific/boys-too-forgotten-stories-human-trafficking/. Accessed August 30, 2015.

Grossman, N. (2015). *Thailand's sustainable development sourcebook*. Bangkok: Editions Didier Miller.

Hodal, K., Lawrence, F. & Kelly, C. (2014). *Revealed: Asian slave labour producing prawns for supermarkets in US, UK*. (online) The Guardian. Available at: https://www.theguardian.com/ global-development/2014/jun/10/supermarket-prawns-thailand-produced-slave-labour. Accessed July 8, 2016.

Hodal, K. & Kelly, A. (2016). Thailand's improved status in US human trafficking report sparks fury (online) *The Guardian*. Available at: https://www.theguardian.com/global-development/ 2016/jun/30/thailand-us-trafficking-in-persons-report-2016-fury. Accessed August 4, 2016.

International Labour Organization. (2014). *Global wage report 2014/15: Asia and the pacific supplement*. (online) Bangkok: International Labour Organization, p. 2. Available at: http:// www.ilo.org/wcmsp5/groups/public/—asia/—ro-bangkok/—sro-bangkok/documents/ publication/wcms_325219.pdf. Accessed July 4, 2016.

International Labor Rights Forum and the Migrant Workers Rights Network. (2016). *Building a Rights Culture: How workers can lead sustainable change in Thailand's seafood processing sector*. (online) Washington, DC, pp. 23–26. Available at: http://www.laborrights.org/sites/ default/files/publications/MWRNreportonline.pdf. Accessed June 6, 2016.

Jayagupta, R. (2009). The Thai government's repatriation and reintegration programmes: Responding to trafficked female commercial sex workers from the greater mekong subregion. *International Migration*, (online) 47(2), pp. 227–253. Available at: http://onlinelibrary.wiley. com/doi/10.1111/j.1468-2435.2008.00498.x/abstract. Accessed July 8, 2016.

Jeffrey, L.A. (2002). *Sex and borders. Gender, national identity, and prostitution policy in Thailand*. Silkworm Books: Chiang Mai.

Jones, S. (2010). The invisible man: The conscious neglect of men and boys in the war on human trafficking. *Utah Law Review*, (online) 4,1143–1188. Available at: http://epubs.utah.edu/index. php/ulr/article/view/484/352. Accessed June 20, 2016.

Long, L. (2004). Anthropological perspectives on the trafficking of women for sexual exploitation. *International Migration*, (online) 42(1), 5–31. Available at: http://onlinelibrary.wiley.com/doi/ 10.1111/j.0020-7985.2004.00272.x/full. Accessed June 3 2016.

Piper, N. (2005). A problem by a different name? A review of research on trafficking in south-east asia and oceania. *International Migration*, (online) 43(1-2), 203–233. Available at: http:// onlinelibrary.wiley.com/doi/10.1111/j.0020-7985.2005.00318.x/epdf. Accessed June 6, 2016.

Rafferty, Y. (2013). Child trafficking and commercial sexual exploitation: A review of promising prevention policies and programs. *American Journal of Orthopsychiatry*, (online) 83(4), 559–575. Available at: http://onlinelibrary.wiley.com/doi/10.1111/ajop.12056/abstract. Accessed July 5, 2016.

Servaes, F. (2015). Sex trafficking issues in Thailand: The case of Urban Light. paper presented at International Conference "Communication/Culture and the Sustainable Development Goals. Challenges for a new generation (CCSDG)", Chiang Mai University: Thailand, 17–21 December 2015.

Thailand Global Monitoring Status of Action Against Commercial Sexual Exploitation of Children (Publication). (2011) ECPAT International. http://resources.ecpat.net/A4A_2005/PDF/EAP/A4A2011_EAP_Thailand_FINAL.pdf. Accessed July 25, 2016.

Transparency International (2013). Corruption Perception Index 2013. (online) Available at: https://www.transparency.org/cpi2013/results. Accessed August 3, 2016.

Transparency International, (2015). *ASEAN Integrity Community: A Vision for Transparent and Accountable Integration*. (online) Transparency International. Available at: http://www.transparency.org/whatwedo/publication/asean_integrity_community. Accessed July 13, 2016.

U.S. Department of State. (2015). *Trafficking In Persons Report 2015*. 330–334.

U.S. Department of State. (2016). *Trafficking in Persons Report 2016*. 363–368.

Whiting, A. (2015). *Tech-savvy sex traffickers stay ahead of authorities as they lure teens online*. (online) PhillyVoice. Available at: http://www.phillyvoice.com/tech-savvy-sex-traffickers-stay-ahead-of-authoriti/. Accessed July 13, 2016.

Websites

Transparency.org (2013). *Corruption Perceptions Index 2013*. [online] Available at: https://www.transparency.org/cpi2013/results. Accessed July 2, 2016.

Urbanlight.org. http://www.urban-light.org/.

UNICEF. (2014). *Hidden in plain sight: A statistical analysis of violence against children*. (online) Available at: http://www.unicef.org/publications/index_74865.html. Accessed June 12, 2016.

Chapter 9
Elephants in Tourism. Sustainable and Practical Approaches to Captive Elephant Welfare and Conservation in Thailand

Lisa Servaes

Abstract This chapter analyzes the complexities of the issues surrounding captive elephants within Thailand's tourism industry. It aims to assess the current status and urgency of elephant conservation; presents an overview of the role of captive elephants in tourism; addresses the animal welfare issues in Thailand's tourism industry as presented by scholars and NGOs; and proposes feasible approaches toward sustainable alternatives and solutions.

9.1 Introduction

Thailand has been severely criticized by international animal rights, animal welfare, and wildlife conservationist groups for its use of captive elephants for human entertainment in the tourism industry. These groups, including the international media, have highlighted the issues pertaining to animal welfare within this industry in various reports and news articles. However, reports oftentimes focus only or too vehemently on one aspect of a larger issue which misrepresents the complexities of said issue. There are important environmental, social, economic, and political issues linked to the predicament of elephants in tourism which are not considered (Elephant Nature Foundation n.d.). Realistic and sustainable solutions are necessary to the welfare of captive elephants in tourism and the conservation of the endangered species. This chapter serves to analyze the complexities of the issues surrounding captive elephants within Thailand's tourism industry using scholarly works, NGO and media reports, and published case studies. Its goals are to consider the current status and urgency of elephant conservation; present an overview of the role of captive elephants in tourism; address the animal welfare issues in Thailand's tourism industry as presented by scholars and NGOs; and assess feasible (and infeasible) approaches toward sustainable alternatives and solutions.

The original version of the chapter was revised: Missing author name has been updated. The erratum to the chapter is available at 10.1007/978-981-10-4125-9_10

9.2   An Anthropocentric History

Traditionally, a highly valued species that has lived alongside Thai people for centuries, the elephant's contributions to Thailand's history are well documented in Thai religion and art (Cadigan 2016). Elephants played an integral role in the functioning of everyday life in Thailand for transportation, agricultural activities, and warfare until the industrial revolution (Cohen 2008: 163–169). Thereafter, working elephants in Thailand were mostly used in the logging industry until the 1989 Logging Ban was implemented by the Royal Thai Government. This statewide ban on commercial forestry effectively rendered 70% of Thailand's working elephants without work practically overnight. Stripped off their incomes and with few alternatives, many mahouts (the riders and keepers of elephants) were forced to use their elephants for illegal activities by continuing their logging work, rendering the ban ineffective on an ecological level (Elephant Nature Foundation, n.d.; Laohachaiboon 2010). The other alternative to working in the illegal logging industry was begging in the busy streets of Thailand's major tourist cities which posed serious dangers to the mahout, the elephant, and tourists (Winkler and Creative 2015: 5). The proposed solution to what became known as Thailand's "elephant problem" was to use elephants in Thailand's tourism industry. An initiative was launched by the Thai Elephant Conservation Center, a government authority, to promote the use of elephants in tourism as a means to provide sustainable employment for elephants and mahouts (Laohachaiboon 2010). Since their traditional purposes had become replaced by machinery, elephants had no other option but to enter the tourism industry. The elephant, recognized as the symbolic animal of ancient Siam (Cohen 2008: 9), was effectively transformed into a tourist attraction.

Wildlife tourism encompasses the entire realm of tourism associated with viewing and encountering wildlife ranging from captive, semi-captive, and non-captive (wild) settings. Its definition includes a variety of interactions from passive observation to feeding and/or touching the species (Newsome et al. 2005). Cohen (2008: 31) notes that the majority of tourists in Thailand do not prioritize "authentic" encounters with wild animals as one would in a non-captive setting. Instead, most tourists seek presentations of animals in semi-captive settings such as elephant camps, and completely captive settings, as in animal entertainment shows. The popularity of semi-captive and captive settings mandates the animals' removal from its natural habitat and marks the prevailing anthropocentric view held by the wildlife industry. An estimated 2300 of Thailand's captive elephants are employed in Thailand's tourism industry. About 135 elephant camps and other tourist establishments are located around major foreign tourism centers such as Bangkok, Chiang Mai, Pattaya, and Phuket (Kontogeorgopoulos 2009a, b; Cohen 2008: 164). The elephant has become the most widely used captive wildlife in Thailand's tourism industry (Cohen 2008). Elephant camps vary in their size and the activities that they offer, although bigger camps tend to offer elephant performances and smaller camps mainly provide elephant rides (Cohen 2008; Winkler and Creative 2015).

Elephants in the tourism industry generally provide three services: direct interaction with tourists, entertainment, and demonstration of obedience, skills, and intelligence (Kontogeorgopoulos 2009a, b). In entertainment shows, elephants are trained to perform various tricks such as playing football, walking on a tightrope, throwing darts, sitting on stools, walking on their hind legs, painting pictures, dancing, playing music, or even flushing a toilet (Cohen 2015; Kontogeorgopoulos 2009a, b). These shows are tailored for mass tourists on brief vacations in the country and take no special effort to find because they are so widespread (Cohen 2008: 155).

9.3 Welfare Issues

The demand for wildlife tourism is growing, which has raised important ethical issues concerning animal welfare and how appropriate and acceptable use of animals for human entertainment is defined (Duffy and Moore 2011: 596). The use of captive elephants in Thailand's tourism industry, and captive wildlife in general, has long been condemned by international and national animal rights and welfare NGOs over concerns for their treatment (Nijman 2014). World Animal Protection (WAP), an international animal welfare organization, uses guidelines identified in 2003 by the United Kingdom's Farm Animal Welfare Council named the "Five freedoms for animal welfare" as primary standards for animals used for tourism purposes. The list follows that animals must have (World Animal Protection 2010):

1. Freedom from hunger and thirst
2. Freedom from discomfort
3. Freedom from pain, injury, and disease
4. Freedom to express normal behavior
5. Freedom from fear and distress

The welfare of captive wildlife is strongly dictated by how closely their living conditions resemble their natural wild habitats, although WAP is adamant that the needs of wild animals can only be fully met in the wild (World Animal Protection 2010). In 2010, WAP compiled a survey on elephants used for entertainment in Thailand. Elephant venues throughout Thailand were assessed based on animal welfare conditions and given a rating as commendable, inadequate, or severely inadequate. Out of the elephant venues surveyed, 80% were classed as severely inadequate, 15% were classed as inadequate, and only 5% of the elephant venues in Thailand were given a commendable rating. The main welfare issues were as follows: the severe control of the animals' freedom of movement by chaining or containment to small cages, limited opportunity for social interaction, participation in stressful and physically demanding show activities, insufficient or nonexistent veterinary care, and inadequate nutrition (World Animal Protection 2010).

A common misconception about captive elephants used in Thailand's tourism industry is that the elephant is a domesticated animal. Domestication entails that an animal differs significantly in anatomy and character in contrast to its wild counterpart, due to generations of selective breeding by human beings (Lair 1997). Despite having been used by humans for thousands of years, elephants have never undergone the domestication process and are forced into submission by their human captors. They remain as one of the most dangerous wild animals to handle despite their perceived calm behavior while carrying humans on their backs and during entertainment shows. Their calm behavior is a product of their forced submission—a breaking-in process that elephant calves are subjected to at around three years old (World Animal Protection). In the past, breaking-in procedures were conducted in secrecy and surrounded by complex magical rituals by ethnic groups, such as the Karen in northern Thailand. The ceremony is called "paah jaan" and is intended to sever the strong bond between the calf and its mother in order to transfer it from the mother to the calf's mahout. The will of the calf is broken during this process, and the calf is made to submit to the mahout (Cohen 2015: 163–169).

Thoswan Devakul (cited in Cohen 2015: 163–169), a wildlife photographer, described the extreme suffering that the elephant calf was subjected to during the breaking-in process. The calf is firstly separated from its mother into an enclosed wooden cage. The calf is then hit with the mahout's nailed bamboo stick on his legs and trunk while he cries helplessly for his mother. Devakul noted that holy water was poured on the calf's head, which is believed to tame him, but an iron hook to the head is followed soon after. Visiting the cage later on, he described how the elephant's ears were "bathed with blood from wounds made by the hook." The elephant had tears in his eyes and he had been chained by the neck and legs. The length of this process depends on how quickly the calf submits to the mahout. Devakul concluded: "What makes the elephants so tame is not the food and care they receive from the mahouts, but the fear from the pain from iron hooks and the suffering they experience from being separated from their mothers before they are naturally ready" (Cohen 2015: 163–169).

The forced submission of elephants in the tourism industry is the premise to arguments that animal welfare groups like WAP make to discourage elephant trekking rides and elephants shows. Elephants are forced to perform and give rides to tourists out of threat and fear of the mahout's hook, which can cause painful puncture wounds when used forcefully and inappropriately (Kontogeorgopoulos 2009a, b: 6). In addition to the welfare issues outlined by WAP's survey, their natural behaviors are also suppressed by cruel treatment like during the "musth" period, an annual phase of increased testosterone production in male elephants when they display unpredictable and aggressive behavior. This period is both inconvenient and dangerous to mahouts, so it is dealt with by chaining the male elephant in isolation for the entire musth period which can last from three weeks to a few months (World Animal Protection 2010).

9.4 Elephant Tourism and Conservation

Thailand's total elephant population has drastically decreased within the past century from a total population of approximately 100,000 in 1900 to a significantly lower number of 4450 in 1990 (Kontogeorgopoulos 2009a, b). Today, an estimated 1000–1500 of Thailand's elephants remain in the wild, most living in protected areas such as Khao Yai National Park and Huay Keng Wildlife Sanctuary (Elephant Nature Foundation n.d.; Kontogeorgopoulos 2009a, b). However, the survival of the Asian elephant is threatened by the poaching that occurs in Thailand's national parks and wildlife sanctuaries and the illegal live elephant trade in Myanmar, both of which are fueled by the demand for baby elephants in Thailand's tourism industry (Cohen 2008: 10). The tourism industry demands elephant calves because they have a greater appeal to tourists who will pay more to have close contact with baby elephants. This is a new trend as traditionally adult elephants were sought after for wild capture since they could be trained and put to work (Nijman 2014: 8). Most baby elephant poaching within Thailand occurs in the wilderness areas of western Thailand, along the border with Burma and particularly in Kaeng Krachan National Park (Cohen 2015: 174). Capturing them from the wild is preferred over breeding them in captivity due to their long gestation period, low birth rates in captivity, and the ease at which they can be trafficked across the Myanmar border (Nijman 2014: 9).

The capture of wild elephants to supply the tourist trade in Thailand is now being recognized as a potentially significant threat to the entire species (Nijman 2014: 8). It has been estimated that as many as 75% of captive adult elephants used for tourism entertainment have been taken directly from the wild (World Animal Protection 2010). Non-captive wild elephants in Thailand are governed under the Wild Animal Reservation and Protection Act 1992 (WARPA) which protects the hunting, killing, and trading of wild elephants or wild elephant parts. Perpetrators will receive a fine of USD 1330 and/or up to four years in prison. Nijman (2014: 6) rejects these penalties as sufficient deterrents to elephant traffickers, as the current value of live elephants on the market is at least 25 times greater than the fine. Elephant calf poaching causes at least one or more elephant fatalities. Poachers often kill the calf's mother and sometimes more members of the herd as they will fiercely protect the calves against an intruder (Cohen 2015: 175). Unfortunately, there are more captive elephants in Thailand than there are wild, and the continuation of illegal elephant poaching is possible through a loophole within the registration policy of the Draught Animal Act 1939 which covers captive elephants and gives them the same legal rights as livestock—a reminder of Thailand's traditional ideology regarding elephants (Cohen 2015; Nijman 2014). The system only requires that animals over the age of eight be registered with the appropriate government department, meaning that elephants younger than eight years old are easily illegally poached from the wild and snuck in among the captive elephant population and then registered as a "domestic" elephant (Nijman 2014: 6). This loophole prevents the necessary protection of wild elephants and is extremely counterproductive to wild elephant conservation efforts. It is another detriment to the elephant tourism industry.

9.5 Plight of the Mahout

Since the cultural status of elephants has decreased and replaced with their economic value, the cultural status of mahouts has also considerably dropped. Traditionally, mahouts endured years of apprenticeship, economic hardship, and physical toll before becoming qualified mahouts (Nijman 2014: 11). The once noble profession is now associated with low levels of status and self-esteem, a further detriment to the already socially and economically marginalized position of indigenous hill tribe minority communities to which many mahouts belong (Kontogeorgopoulos 2009a, b: 7–8). Ethnic minority refugees from Myanmar are also commonly employed as mahouts (Cadigan 2016). There are important considerations around how mahouts will earn a living without captive elephants (Duffy and Moore 2011: 594). The solution is not simple when those employed as mahouts are in many cases undocumented migrant workers from neighboring countries who simply do not have an alternative. These workers are paid below minimum wages (Cadigan 2016), effectively bonded to their employers and at risk of rights violations from government authorities (Human Rights Watch 2010).

Mahouts have an instrumental role in individual elephant welfare as they are the direct handlers of elephants in the tourism industry. The mahout profession traditionally required a lifelong interest in and dedication to elephants, resulting in a mahout's deep commitment to the elephant in their care. Today, most mahouts are poor, uneducated young men who will easily abandon their elephants when better paying work comes along because they never intended to commit to their elephants long-term care (Lair 1997). This dynamic hurts camp owners, elephants, and mahouts. Camp owners will require additional resources for training, elephants will suffer when mahouts with insufficient skills and experience resort to rough methods to control them, and mahouts with poor familiarity of that elephant's temperament leads to fatalities (Kontogeorgopoulos 2009a, b: 8). Mahouts earn 3000–5000 baht ($85–141) monthly salary and can earn between 200 and 1000 baht (5.5–28 $) per day in tips from tourists which, as extra income, reduces the economic incentive for mahouts to seek supplemental employment outside the camp and thus increases the time spent between mahouts and their elephants. More income also means better veterinary care, greater quantity of food, and a better range of food sources (Kontogeorgopoulos 2009a, b: 10–11).

9.6 Role of Western NGOs

Western NGOs have been at the forefront of raising awareness about the animal welfare issues in Thailand's tourism industry. However, their approaches have been criticized by scholars for their cultural insensitivity and simplification of the complexity of Thailand's captive elephant issue (Cohen 2013; Duffy and Moore 2011; Kontogeorgopoulos 2009a, b). In 2002, People for the Ethical Treatment of

Animals (PETA) launched the issue of animal welfare in Thailand's tourism industry when they received a video of a young elephant enduring the "paah jaan" ceremony. After Thai officials did not act on the matter, PETA launched an "Abusive Thailand: Elephant Cruelty" campaign in Australia (Cohen 2015: 170) and demanded that the practice be stopped. PETA then called for a total boycott of tourism to Thailand until the Thai government enacted laws prohibiting the use of elephants for commercial entertainment purposes. PETA claimed that the elephants were tortured to make them perform tricks in circuses and shows as well as providing rides for tourists (Cohen 2015; Duffy and Moore 2011; Kontogeorgopoulos 2009a, b). Elephants had become an integral part of Thailand's tourism industry by 2002, making PETA's campaign a direct threat to the industry (Cohen 2015). Cohen (2015: 176) explained that "when outsider activists introduced the discourse of animal rights and welfare into the Thai public sphere, they met with incomprehension by the representatives of the elephant training establishments, who claimed that the alleged 'cruelty' in taming young elephants is part of an ancient tradition, even as the higher authorities sought, somewhat disingenuously, to deny the exercise of any cruelty in the taming of young elephants and condemned the allegations of cruelty as vicious recriminations."

Kontogeorgopoulos (2009a, b: 16) argues that "Domestic and particularly foreign NGO's interested in elephants or animal rights more generally have hurt their cause by unnecessarily antagonizing elephant camps with the strident tone and inflexibility of their criticisms," while Cohen (2013: 281) asserts that international welfare NGOs must be careful in their impositions of Western ethical standards upon a non-Western culture. Western animal rights and welfare NGOs view Thailand's captive elephants as having intrinsic value, whereas Thai culture views captive elephants from a utilitarian perspective (Duffy and Moore 2011; Cohen 2008). Cohen (2013: 278) argued that "the concept of 'animal abuse,' as broadly conceived in the contemporary West, is external to Buddhist attitudes to animals in present-day Thailand; hence the criticism that animals are 'abused' is incomprehensible to many Thais."

Western NGOs call for tourists not to support any forms of animal entertainment in the form of shows or riding opportunities with the end goal that captive wildlife will be released back into the wild where "they belong" or to be kept in sanctuaries, free from their work in entertaining tourists. The problem with this is that captive elephants are not "wild," and the great majority would not be able to adapt to a wild habitat. Additionally, there is insufficient space to accommodate all of Thailand's captive elephants in its existing national (Duffy and Moore 2011: 294). Thailand's forest land reduced from 80% in 1957 to less than 20% in 1992, effectively destroying wild elephant habitat (Schliesinger 2010: 19). This is a barrier, among several others, that many animal welfare NGOs fail to acknowledge in their "keep animals in the wild" rhetoric (Duffy and Moore 2011). Duffy and Moore (2011: 598) argue against the idea that all elephants should be in the wild because "it fails to constructively engage with the everyday realities of elephant conservation in Thailand."

9.7 Toward Sustainable Solutions

Though the use of elephants in any form of tourism-related work is condemned by some animal rights activists to be demeaning and unnatural, Kontogeorgopoulos (2009a, b: 3) argues that given the current absence of viable alternatives, tourism contributes to the welfare of working elephants in Thailand in "optimal, albeit imperfect ways" and that the promotion of more tourism, not less, is the most realistic way to improve their welfare. It is an imperfect solution from an animal rights perspective to keep elephants in the tourism industry, but it is argued that their value is based on market forces, and a demand for working elephants leads to better care and protection than would be the case if their values were to plummet (Kontogeorgopoulos 2009a, b: 10).

ASEAN Captive Elephant Working Group, a group of regional elephant specialists, veterinarians, researchers, and conservationists, called for realistic solutions to ensure the sustainable and ethical management of captive elephants in the future in their statement concerning the issue of captive elephant welfare and conservation in Asia (ASEAN Captive Elephant Working Group 2015). Duffy and Moore (2011: 596) further argue that captive working elephants are important to the long-term conservation of elephants in Thailand because without them, the long-term survival of the entire species would be at risk. Therefore, closing down the elephant trekking industry greatly risks the survival of the species. This calls for a prioritization of elephant welfare in the captive elephant population and a prioritization in the conservation of the wild elephant population.

Elephant tourism venues in Thailand do not operate by a concern for elephant conservation or animal welfare, which calls for a transition into more sustainable and ethical practices that ecotourism promotes. Lair (1997) argues that the goal of keeping all of Thailand's captive elephants in sanctuaries will encounter severe obstacles. Namely, they are extremely expensive to start, due to land and infrastructure costs, and need continued funding to maintain. He states that "there still remains a huge monthly overhead in salaries because sanctuaries will succeed for healthy, privately-owned elephants only if the mahouts' earnings within the sanctuary equal what they can make outside." However, several elephant sanctuaries have emerged in Thailand that cater to the tourism industry while prioritizing the welfare and conservation of Thailand's captive elephants by not allowing elephant riding or entertainment shows. A case study on Elephant Nature Park (ENF) in Chiang Mai, an elephant sanctuary catering to the tourism industry, examined how Western-style advocacy was implemented into the context of Thailand to pioneer the adoption of abused domestic elephants and implement sustainable ecotourism in Thailand (Lin 2012). ENF advocates a natural elephant habitat and work-free lifestyle which challenges the traditional ideology surrounding Asian elephants in Thailand. Through soft persuasion and cultural sensitivity, ENF was successful in adapting ecotourism practices to the host environment gradually (Lin 2012: 205). ENF was effective in gradually transforming mahout's old traditions of harsh treatment toward elephants and educating the public (local and tourists) about compromised elephant welfare in tourism entertainment. This type of ecotourism model benefits elephants, mahouts, local villagers, and Thailand's tourism. ENF's success leads to a partnership with the Surin government to build Surin's elephant village project which offers mahouts a steady income and lodging to allow elephants to live their lives free from entertaining tourists (Lin 2012: 207). ENF serves as an example of where Thailand's tourism industry should move toward, although the issue of elephant welfare demands practical implementations in the industry's current elephant camps.

Kontogeorgopoulos (2009a, b: 443) argues that the current anthropocentric camps which allow elephant riding have an important role in the overall welfare of Thailand's domesticated elephants because elephant sanctuaries are difficult to replicate in anything other than an incremental fashion. He states that "critics of the use of elephants in tourism should acknowledge the financial utility and value that anthropocentric camps create for elephants" because elephant riding in his study was ranked as the most important component of tourists' visits to the camp. Due to the competitive business environment of elephant camps in Thailand tourism industry, moving into the direction of not allowing riding at all poses a problem for anthropocentric camps and would likely be detrimental to the welfare of the elephants as opposed to beneficial (Kontogeorgopoulos 2009a, b: 445). He concludes, "It appears that in the short term at least, the welfare of working elephants in Thailand is best served by a combination of continued tourism demand and sustained pressure on camp owners and managers to move camps in a slightly more ecocentric direction" (Kontogeorgopoulos 2009a, b: 445).

On the one hand, tourism supports the welfare of elephants by maintaining their economic value, but on the other hand, their economic value gives incentive to continue poaching wild elephants. This calls for an effective captive elephant management and registration program to identify individual captive elephants and prevent the ability for the inappropriate registration papers to be used for illegally obtained elephants (Lair 1997). Currently, captive elephants are a commodity by law under the Draught Animal Act 1939. Captive elephants and wild elephants should be governed under the Wild Animal Reservation and Protection Act 1992 to give captive elephants more legal rights in efforts for the conservation of the entire species.

The Prevention of Animal Cruelty and Provision of Animal Welfare Act was passed on in November of 2014 as Thailand's first animal welfare law. This law protects domestic pets, farmed animals, working animals, animals kept for entertainment, wild animals in captivity, and any other animals under human care (Geer 2016). The law prohibits owners and carers from treating animals cruelly and providing inadequate living conditions for the animals. It allows police to enter homes and businesses to investigate claims of animal abuse and neglect. Perpetrators can be fined up to THB 40,000 and/or a two-year jail sentence. This law has the ability to protect captive elephants in the tourism industry from cruel treatment and inadequate welfare conditions. However, the law has garnered criticism for being too vague in its definitions of what exactly constitutes "torture" and "cruelty" as the law only described what is not considered to be "cruelty" and leaves what constituted "cruelty" open to interpretation. This leads to questioning of how the law will be enforced and how it can be effective in the case of captive working elephants in the tourism industry (Ambrose 2015).

As the direct handler of elephants and their welfare, mahouts should be required to endure a training program that promotes the welfare of elephants by educating mahouts about their needs. The ASEAN Captive Elephant Working Group (2015) recommends that "modern techniques of animal training should be developed and introduced gradually to adjust or supplement those traditional training and handling practices that cause severe discomfort and suffering."

Western or international NGOs and national NGOs serve an important role in raising awareness for animal welfare in Thailand's tourism industry, educating the public about elephant conservation and raising demands for ecotourism. The cooperation among national NGOs is also crucial for advancing the interests of domesticated elephants in Thailand. Kontogeorgopoulos (2009a, b: 16) notes that there are at least 12 national organizations focusing on the problems faced by domesticated or wild elephants and that there is a lack of coordination and cooperation between these organizations as a result of personal rivalry or divergent beliefs on strategy.

Additionally, the Tourism Authority of Thailand (TAT) should endorse elephant camps that are deemed more professional and ethical in terms of human resource management and animal welfare. This would add additional incentive for camps to operate in ways beneficial to elephants (Kontogeorgopoulos 2009a, b: 16–17) and would take the necessary steps into the direction of ecotourism.

9.8   Concluding Remarks

From an animal rights perspective, the use of captive elephants or any animal in tourism should be abolished, but this is impractical given the current situation: There are more captive elephants than wild, captive elephants cannot adapt to the wild, and there is not enough protected land for all of Thailand's captive elephants. In addition, elephants have a utilitarian value in Thai culture meaning that their welfare is determined by their economic value and their economic value is dependent on their demand in tourism. A compromise has to be made in order to prioritize the welfare of elephants in captivity and to prioritize the conservation of the wild population. A solution that maintains their economic value in the tourism industry and prioritizes their welfare are ecotourist sanctuaries where elephants are allowed to display their natural behaviors without giving rides to tourists or performing in entertainment shows.

This is the end goal for captive elephants in Thailand. However, actions that need to be taken toward those goals need to be practical given the current situation. Immediate actions must be taken by the government to govern captive elephants and wild elephants under the same law and to prevent the continuation of the live elephant trade; all captive elephants must be registered and accounted for to monitor for illegal activities; welfare laws must be made clear and rigorously enforced; mahouts must endure training to promote elephant welfare; and ecotourism should be promoted by The TAT. This complex issue certainly has no definite solutions; however, what is certain is that the Thai government, NGOs, and international bodies must work together to ensure that the fate of the Asian elephant does not include extinction.

References

Ambrose, D. (2015). *Tourist dollars feed wild animal trade in Thailand* (online) Aljazeera.com. Available at: http://www.aljazeera.com/indepth/features/2015/06/tourist-dollars-feed-wild-animal-trade-thailand-150624085459227.html. Accessed July 23, 2016.

ASEAN Captive Elephant Working Group. (2015). *Addressing a giant problem in Southeast Asia*. ASEAN Captive Elephant Working Group.

Cadigan, H. (2016). The human cost of elephant tourism. *The Atlantic* (online). Available at: http://www.theatlantic.com/science/archive/2016/05/elephants-tourism-thailand/483138/. Accessed July 18, 2016.

Cohen, E. (2008). *Explorations in Thai tourism* (pp. 135–178). Emerald: Bingley.

Cohen, E. (2013). "Buddhist Compassion" and "Animal Abuse" in Thailand's tiger temple. *Society & Animals, 21*(3), 266–283.

Cohen, E. (2015). Young elephants in Thai Tourism: A fatal attraction. In K. Markwell (Ed.), *Animals and tourism: Understanding diverse relationships* (1st ed., pp. 163–177). Bristol: Channel View Publications.

Duffy, R., & Moore, L. (2011). Global regulations and local practices: The politics and governance of animal welfare in elephant tourism. *Journal of Sustainable Tourism, 19*(4–5), 589–604.

Elephant Nature Foundation. (n.d.). *An overview of the captive elephant situation in Thailand*. Surin: Elephant Nature Foundation.

Geer, A. (2016). *Thailand creates first animal welfare law, but is it too vague? |Care2 Causes* (online) Care2.com. Available at: http://www.care2.com/causes/thailand-creates-first-animal-welfare-law-but-is-it-too-vague.html. Accessed July 23, 2016.

Human Rights Watch. (2010). *From the Tiger to the Crocodile* (online) Human rights watch. Available at: https://www.hrw.org/report/2010/02/23/tiger-crocodile/abuse-migrant-workers-thailand. Accessed July 18, 2016.

Kontogeorgopoulos, N. (2009a). The role of tourism in elephant welfare in Northern Thailand. *Journal of Tourism, 10*(2), 1–19.

Kontogeorgopoulos, N. (2009b). Wildlife tourism in semi-captive settings: a case study of elephant camps in northern Thailand. *Current Issues in Tourism* (online) *12*(5–6), 429–449. Available at: http://soundideas.pugetsound.edu/cgi/viewcontent.cgi?article=1072&context=faculty_pubs. Accessed July 27, 2016.

Lair, R. (1997). *Gone astray*. Bangkok, Thailand: FAO Regional Office for Asia and the Pacific.

Laohachaiboon, S. (2010). *Conservation for Whom? Elephant Conservation and Elephant Conservationists in Thailand*. Southeast Asian Studies, 48(1).

Lin, T. (2012). Cross-platform framing and cross-cultural adaptation: examining elephant conservation in Thailand. *Environmental Communication, 6*(2), 193–211.

Newsome, D., Dowling, R., & Moore, S. (2005). *Wildlife tourism* (p. 2). Clevedon: Channel View Publications.

Nijman, V. (2014). *An assessment of the live elephant trade in Thailand* (pp. 1–38). Cambridge: TRAFFIC.

Schliesinger, J. (2010). *Elephants in Thailand*. Bangkok: White Lotus.

Winkler, B., & Creative, P. (2015). *Creating awareness & respect for elephants*. Tourism cares (online) Mahouts Elephant Foundation. Available at: http://www.mahouts.co.uk/image/data/blog/tourismcares.pdf. Accessed July 24, 2016.

World Animal Protection. (2010). *Wildlife on a Tightrope* (online) London: World Animal Protection, 1–44. Available at: http://www.worldanimalprotection.org/sites/default/files/int_files/wildlife-on-a-tightrope-thailand.pdf. Accessed July 20, 2016

Erratum to: Culture and Communication in Thailand

Patchanee Malikhao

Erratum to:
P. Malikhao, *Culture and Communication in Thailand*, Communication, Culture and Change in Asia, DOI 10.1007/978-981-10-4125-9

The original version of the book was inadvertently published without implementing the special instruction to include the author names 'Jan Servaes', 'Fiona Servaes' and 'Lisa Servaes' in Chapters 4, 8 and 9, respectively, which have to be now included. The erratum book has been updated with the changes.

The updated original online version of this book can be found at
https://doi.org/10.1007/978-981-10-4125-9_4
https://doi.org/10.1007/978-981-10-4125-9_8
https://doi.org/10.1007/978-981-10-4125-9_9

© Springer Nature Singapore Pte Ltd. 2017   E1
P. Malikhao, *Culture and Communication in Thailand*, Communication,
Culture and Change in Asia 3, DOI 10.1007/978-981-10-4125-9_10

Index

A
Achievement-task orientation, 62
Advertisement, 13, 24, 45, 43, 103
Amnaj/Power, 99
Amoral power, 54, 56, 65
Anatta/Not-self, 18, 106
Anicca, 25, 28, 55, 106
Animal welfare, 127, 129, 132–136
Animism, 1, 6–8, 13, 18, 41, 53, 55–57, 60, 63
Animistic beliefs, 1, 3, 7, 13
Archaic globalization, 2, 3, 7, 74
ASEAN Economic Community (AEC), 62
Astrology, 1, 6, 8, 13
Atta/Self, 18, 25, 26, 28, 106

B
Barami, 65
Bhikkhuni, 38, 39, 42
Bhikku, 8, 17, 21, 105, 110
Biopiracy, 100
Biotech industry, 100
Blogs, 21–24
Brahmanism, 1, 3, 6, 8, 9, 13, 29, 39, 55, 63
Brahmavihara, 111
Buddha, 1, 5, 8, 9, 17, 18, 22, 25, 27–30, 38, 49, 53, 55, 105–109
Buddhadasa Bhikku, 18
Buddha-Dhamma, 7, 9
Buddhism, 17, 18, 26, 28, 30–32, 39, 42, 54–56, 63, 113
Bun-khun, 58, 59

C
Chatukham-Ramathep, 11
Chemical fertilizer, 99, 100
Child trafficking, 120
Coerseductive, 51
Collectivism, 6
Compassion, 28, 94, 101, 109–111, 114

Compost, 97, 99
Conflict avoidance, 112
Consumerism, 7, 8, 13, 18, 33, 99, 103, 113
Contemporary globalization, 1, 3, 6, 13, 17, 21, 29, 40, 74–77, 82, 90, 100, 110
Creolization, 2
Cults, 3, 4, 7–13, 18
Cultural products, 3, 10, 49, 103
Cultural relativity, 72, 73, 82

D
Dana, 30, 95
Dependency paradigm, 93
Dependent origination, 18, 26, 107, 108
Developed countries, 93, 101
Developing countries, 89, 90, 92, 101
Dhammakaya, 8, 17–23
Dhammayuti Order, 31
Digital communication, 17, 19, 20, 71, 85
Digital divide, 75
Digital literacy, 103
Ditthi, 26, 92, 107
Domestic violence, 37, 39

E
Ecology, 74, 78, 91
Economic growth, 77, 90–92
Economic sufficiency scheme, 98
Ecotourism, 77, 78, 83–85, 134
Ego-orientation, 62, 64
Elephant domestication, 130
Environmental sustainability, 91
Equanimity, 101, 106, 111, 113
Ethnocentrism, 72, 76, 77, 111, 113
Evolvability, 78, 91

F
Face, 43, 49, 60, 64, 80, 83, 106, 111, 122
Facebook, 12, 21–23, 75, 82–84, 104, 119

© Springer Nature Singapore Pte Ltd. 2017
P. Malikhao, *Culture and Communication in Thailand*, Communication,
Culture and Change in Asia 3, DOI 10.1007/978-981-10-4125-9

Femininity, 38, 39, 62
Five aggregates (Panca-khandha), 26
Five Freedoms for Animal Welfare, 129
Five hindrances, 105
Fortune-teller, 60
Four Noble Truth, 108
Fun-pleasure orientation, 62–64

G
Genetically Modified Organisms (GMO), 91,
    100
Genetic engineering, 100
Globalization, 1–5, 7, 12, 71, 72, 74, 90, 93,
    94, 103, 104
Global warming, 89, 90
Grateful relationship orientation, 62, 63
Greenhouse effect, 90

H
Hashtags, 21–23
High culture, 10
Hindu-Brahmanism, 4
Hindu gods, 1, 3, 13
Human rights, 72, 91, 122, 132
Human trafficking, 37, 90, 92, 117–123
Hybridization, 1–3, 5, 7, 17, 33
Hybridized culture, 29
Hype, 11

I
Identity, 9, 10, 12, 25, 49, 51, 75, 76, 83
Individualism, 6, 12, 13, 61, 83, 104, 112
Infosphere, 112
Insight meditation, 28, 106
Instagram, 17, 75, 83, 119
Instant gratification, 29, 83
Interdependence orientation, 62–64
Internal colonization, 5
Itthiphon, 65, 66

K
Karma, 9, 54–57, 62
Kilesa, 18, 26, 28, 54, 107, 111
King Chulalongkorn, 1, 5, 13, 17, 50
King Chulalongkorn cult, 9
King Vajiravudh, 9, 50
Kusala-kammapatha (the wholesome course of
    action), 113

L
Lankawong, 1, 5
Life style, 29
Look Thep, 11

Loving kindness, 101, 111, 113, 114
Low culture, 10

M
Magic monks, 7–9, 29, 53
Mahanikaya Order, 31, 32
Mahayana Buddhism, 28
Male supremacy, 39, 40
Masculinity, 38, 45, 62
Matter/Rupa, 26, 106
Media capital, 11, 75, 103
Mediation, 9, 11, 75
Mediatization, 11–13, 29, 75, 83, 103, 104
Merit-making, 6, 30
Mindful communication, 103, 104, 109, 110,
    114
Mindful journalism, 111, 114
Mindfulness, 103, 105, 106, 114
Mindfulness of dhammas, 105
Mindfulness of the body, 105
Mindfulness of the feeling, 105
Modernization paradigm, 7, 83, 84, 90–93
Moral kindness, 52, 54, 56
Multiplicity paradigm, 77, 84, 89, 93, 101

N
Networking, 32, 75, 83, 85, 89, 93, 94, 100,
    101, 104
New individualism, 12, 83, 104
New media, 10, 12, 75, 103, 104
Nibbana (Pali)/Nirvana (Sanskrit), 18, 109
Non-renewable resources, 91
Not-self, 106

O
Online social communication, 73

P
Panna, 94
Participatory decision making, 90
Paticcasamuppada, 26, 107, 108
Patriarchy, 4, 8, 42, 83
Payutto, P.A, 17, 18, 21, 26, 91
People for the Ethical Treatment of Animals
    (PETA), 133
Phi/spirit/ghost, 3
Phra-dej, 56, 65
Phra-khun, 56, 65
Polygamy, 4, 40
Pop culture, xxiv
Popular artifacts, 10
Popular culture, 10, 75
Popular rituals, 10

Postcolonial globalization, 2, 3
Power distance, 62
Proto-globalization, 1, 2, 40, 74

R
Rape, 37, 43, 45
Rattanakosin, 5
Rejoice, 101
Religio-psychical orientation, 62
Religious culture, 3, 9, 13, 21, 29, 30, 33
Renewable resources
Rest and recreation (R&R), 71

S
Samadhi, 94, 106
Santi-Asoke, 8, 18, 32
Secular, 6, 55
Seed collection, 97, 100, 101
Seed conservation, 89, 100
Self-reflexivity, 101, 103
Self-reliance, 18, 77, 89, 93, 96–98, 100
Sex education, 42, 46
Sex trafficking, 117–120
Sexual exploitation, 119, 120
Sila, 94
Snapchat, 119
Social media, 1, 13, 17, 20, 45, 79, 101, 103, 114, 119
Social network, 85, 113
Social networking sites, 12
Social responsibility, 103
Social smoothing orientation, 62
Social values, 3
Spirit house, 53
Stereotype, 80, 119
Suffering, 4, 18, 27, 55, 98, 99, 101, 104–106, 108–110, 113, 130, 136
Sunyata, 18, 109
Supernatural power, 8, 28, 53

Sustainability, 18, 71, 73, 77, 84, 91, 93, 99, 101
Sustainable development, 71, 72, 77, 89, 91, 93, 96, 112
Sustainable society, 91
Sustainable yield, 91
Symbolic representation, 20, 29–31, 49, 51, 83

T
Thai hybridized religious culture, 25
Thai worldview, 45, 53, 56, 65
The internet, 6, 10, 12, 30, 74, 101, 104
The nirvana, 30, 54
The Paticca Samuppada model, 18
The Precept Five, 6, 110
Theravada Buddhism, 1, 4, 13, 28, 29, 31
The sakdina system, 40, 51, 52
The Sangha, 1, 3, 5, 7, 8, 13, 25, 31, 32
Tourism Authority of Thailand (TAT), 72, 84
Twitter, 12, 21–24, 83, 104, 119

U
Uncertainty avoidance, 62
Underdeveloped countries, 77
Urbanization, 7, 45, 92

V
Value system, 8, 29, 49, 51, 62, 76
Village culture, 4, 8, 52

W
Westernization, 2, 6, 7, 59, 66, 90
Worldview, 3, 8–10, 20, 29, 30, 37, 45, 55, 63, 76, 90, 103

Z
Zero-dollar tourism, 84

The manufacturer's authorised representative in the EU is Springer
Nature Customer Service Centre GmbH, Europaplatz 3, 69115 Heidelberg,
Germany. If you have any concerns regarding our products, please
contact ProductSafety@springernature.com

Printed and bound by CPI Group (UK) Ltd, Croydon, CR0 4YY

29/04/2026

02099460-0016